Nonclinical Study Contracting and Monitoring: A Practical Guide

Nonclinical Study Contracting and Monitoring: A Practical Guide

Edited by

William F. Salminen, PhD, DABT, PMP
PAREXEL International, Sarasota, FL, USA

Joe M. Fowler, BS, RQAP-GLP
Director of the Quality Assurance Unit, National Center for
Toxicological Research, FDA, Jefferson, AR, USA

James Greenhaw, BS, LAT
Center of Excellence for Hepatotoxicity, National Center for
Toxicological Research, FDA, Jefferson, AR, USA

AMSTERDAM • BOSTON • HEIDELBERG • LONDON
NEW YORK • OXFORD • PARIS • SAN DIEGO
SAN FRANCISCO • SINGAPORE • SYDNEY • TOKYO

Academic Press is an Imprint of Elsevier

Academic Press is an imprint of Elsevier
32 Jamestown Road, London NW1 7BY, UK
225 Wyman Street, Waltham, MA 02451, USA
525 B Street, Suite 1800, San Diego, CA 92101-4495, USA

Notice
No responsibility is assumed by the publisher for any injury and/or damage to persons
or property as a matter of products liability, negligence or otherwise, or from any use or
operation of any methods, products, instructions or ideas contained in the material herein.
Because of rapid advances in the medical sciences, in particular, independent verification
of diagnoses and drug dosages should be made

British Library Cataloguing-in-Publication Data
A catalogue record for this book is available from the British Library

Library of Congress Cataloging-in-Publication Data
A catalog record for this book is available from the Library of Congress

ISBN: 978-0-12-397829-5

For information on all Academic Press publications
visit our website at www.store.elsevier.com

Typeset by TNQ Books and Journals

List of Contributors   xi

1. Introduction   1

GLPs and Nonclinical Studies   4
CROs and Nonclinical Studies   6
Study Directors   8
Examples of Study Issues   9
   General Study Issues   9
   Protocol   10
   Test Article   12
   In-Life   13
   Necropsy   15
   Reporting   16
Conclusion   17

2. Good Laboratory Practices   19

US FDA (21 CFR Part 58) and OECD GLPs   22
Subpart A – General Provisions   23
   Section 58.3 (Definitions)   23
   Section 58.10 (Applicability to Studies Performed Under
   Grants and Contracts)   25
   Section 58.15 (Inspection of a Testing Facility)   25
Subpart B – Organization and Personnel   25
   Section 58.29 (Personnel)   26
   Computer Systems (21 CFR Part 11)   26
   Section 58.31 (Testing Facility Management)   26
   Section 58.33 (Study Director)   27
   Section 58.35 (QAU)   27
Subpart C – Facilities   29
   Section 58.41 (General)   29
   Section 58.43 (Animal Care Facilities)   29
   Section 58.45 (Animal Supply Facilities)   30
   Section 58.47 (Facilities for Handling Test and Control Articles) 30
   Section 58.49 (Laboratory Operation Areas)   30
   Section 58.51 (Specimen and Data Storage Facilities)   30
Subpart D – Equipment   30
   Section 58.61 (Equipment Design)   30
   Section 58.63 (Maintenance and Calibration of Equipment)   31
   Computer Systems (21 CFR Part 11)   31

Subpart E – Testing Facilities Operation   31
    Section 58.81 (Standard Operating Procedures [SOPs])   31
    Section 58.83 (Reagents and Solutions)   32
    Section 58.90 (Animal Care)   33
Subpart F – Test and Control Articles   34
    Section 58.105 (Test and Control Article Characterization)   34
    Section 58.107 (Test and Control Article Handling)   34
    Section 58.113 (Mixtures of Articles with Carriers)   35
Subpart G – Protocol for and Conduct of a Nonclinical
Laboratory Study   35
    Section 58.120 (Protocol)   35
    Section 58.130 (Conduct of a Nonclinical Laboratory Study)   36
Subpart J – Records and Reports   37
    Section 58.185 (Reporting of Nonclinical Laboratory Study Results) 37
    Section 58.190 (Storage and Retrieval of Records and Data)   38
    Section 58.195 (Retention of Records)   38
GLP Facility Inspections   40
    European Union (EU) Facility Inspections   42
GLP Auditing Checklist   42

3. Study Design   59

General Study Design Issues   61
    Animal Procurement and Selection   61
    Quarantine and Pre-Study Health Assessment   62
    Animal Identification   63
    Animal Housing   63
    Water and Feed   64
    Environmental Controls   65
    Randomization to Groups   66
    Dosing   67
    In-Life Evaluations   68
    Terminal Procedures   70
    Other Assessments   73
Study Design Checklist   73

4. Animal Welfare   81

The "Guide"   84
Regulations Established by the USDA Under the AWA   92
Potential Conflicts Between the Animal Welfare Requirements,
GLPs, and Other Study Requirements   93
Animal Welfare Checklist   94

5. Laboratory Selection   103

Contacting and Preliminary Screening of a New Laboratory   104
    Meeting with a Sales and/or Technical Representative   105

Visiting and Auditing a New Laboratory 108
Laboratory Selection Checklist 115

6. Project Proposal 121

Detailed Study Outline 122
Price Negotiation 131
Detailed Study Outline Template 132

7. Contracts and Business Ethics 137

Confidentiality 137
Contracts 139
Maintaining Confidentiality During the Study 142
Business Ethics 144
Example of a Confidential Disclosure Agreement 145

8. Study Protocol Preparation, Review, and Approval 149

Writing the First Draft 150
Reviewing the Draft Protocol 151
Title Page 152
Table of Contents 153
Body of Protocol 153
Submission of Study to Regulatory Authorities 155
Test and Control Articles 155
Test System 156
Experimental Design 159
Test and Control Article Administration 159
In-Life Study Evaluations 161
Scheduled Euthanasia 164
Postmortem Evaluations 164
Statistical Methods 166
Study Reports 166
Records and Specimen Retention 166
Compliance with Animal Welfare Regulations 166
Signatures 167
Finalizing the Protocol 169
Changing the Finalized Protocol 170
Protocol Checklist 171

9. Test Article 173

Test and Control Article Synthesis and Sourcing 173
Test and Control Article Characterization 175
Mixtures of Test Articles 176
Test Article and Mixture Receipt, Storage, and Tracking 178
Additional Considerations 179

List of Contributors

Jeffrey Ambroso, PhD, DABT, Dept of Safety Assessment, Glaxo Smith Kline, Research Triangle Park, NC

Amy Babb, BS, National Center for Toxicological Research, FDA, Jefferson, AR

Jeff Carraway, DVM, MS, DACLAM, National Center for Toxicological Research, FDA, Jefferson, AR

Kelly Davis, DVM, National Center for Toxicological Research, FDA, Jefferson, AR

Joe M. Fowler, BS, RQAP-GLP, Director of the Quality Assurance Unit, National Center for Toxicological Research, FDA, Jefferson, AR

Neera Gopee, DVM, PhD, DABT, National Center for Toxicological Research, FDA, Jefferson, AR

James Greenhaw, BS, LAT, Center of Excellence for Hepatotoxicity, National Center for Toxicological Research, FDA, Jefferson, AR

Lady Ashley Groves, BS, BA, National Center for Toxicological Research, FDA, Jefferson, AR

Mark Morse, PhD, DABT, Charles River Laboratories, Spencerville, OH

William Salminen, PhD, DABT, PMP, PAREXEL International, Sarasota, FL

Karen VanLare, BA, RVT, Novartis Animal Health, Greensboro, NC

Xi Yang, PhD, National Center for Toxicological Research, FDA, Jefferson, AR

Chapter 1

Introduction

William F. Salminen PhD, DABT, PMP*, James Greenhaw BS, LAT† and Joe M. Fowler BS, RQAP-GLP†

*PAREXEL International, Sarasota, FL, †National Center for Toxicological Research, FDA, Jefferson, AR

Key Points

- Active involvement with a nonclinical study is needed to ensure a high quality study
- GLPs help ensure data quality and study reconstruction; however, studies can still be inadequate or have GLP deviations
- The Study Director at the laboratory is the person ultimately responsible for the conduct of the study and maintaining GLP compliance
- The Contracting Scientist must establish a solid working relationship with the Study Director and clearly set expectations for not only their interaction but how the study should be run
- Laboratories and Study Directors may excel at certain study functions but may fall short on others

---

The aim of this book is to provide the reader with a hands-on guide for designing, contracting, monitoring, and reporting nonclinical studies. This encompasses both efficacy and safety/toxicology studies. The main focus of this book is on studies that must comply with Good Laboratory Practices (GLPs) for regulatory submissions; however, the guidance is also applicable to studies that do not need to comply with GLPs since general advice on running sound scientific studies is provided. For purposes of reference, the term Contracting Scientist will be used to represent the person at a company that is responsible for designing, contracting, and monitoring a nonclinical study. The Contracting Scientist may be a single person, such as may occur at a smaller company, where he/she has to wear multiple hats. In a larger company, the Contracting Scientist may be comprised of multiple people, each with a specific role to play in various study functions, such as Study Sponsor, Toxicologist, Study Monitor, Test Article Manager, Animal Welfare Officer, etc. Similar to the term Contracting Scientist, the term "Study Contracting and Monitoring" is used to capture all aspects of a study from initial

Nonclinical Study Contracting and Monitoring. http://dx.doi.org/10.1016/B978-0-12-397829-5.00001-6
2013, Published by Elsevier Inc.

laboratory selection through protocol development, conduct of in-life functions, and final reporting of the study. Depending on the organization, these functions may be the responsibility of a single person or the individual functions may be delegated to different people. Finally, the terms "nonclinical" and "preclinical" are both used in this book to describe animal studies. The GLPs technically refer to these studies as "non-clinical" studies; however, the original deliberations that led to the GLPs referred to the studies as "preclinical".[1] That is probably the reason both terms are frequently interchanged. For the purposes of this book, the two terms are considered synonymous.

The idea for this book came out of the frustration of trying to find resources on contracting and receiving sound GLP toxicology study reports from Contract Research Organizations (CROs) and internal company laboratories. Often, a given laboratory would excel in some areas but fall short in others. For example, one CRO was great at conducting the in-life portion of a study and kept on top of everything to ensure that all study functions were conducted like clockwork. However, their data collection procedures were marginal and their reporting of the results was so poor that reports had to be essentially written by the Contracting Scientist. In order to ensure that studies are adequately conducted and reported, companies, particularly larger companies, often have procedures in place for contracting studies with CROs. However, there are often gaps that miss critical aspects of study contracting and monitoring or they do not provide a holistic view of what is involved in running a study from start to finish. The larger the company, the more likely the study contracting and monitoring process is siloed and specific aspects of the process are delegated to various people, which can prevent each individual person from clearly understanding how all the parts fit together. In contrast, in smaller companies, a single Contracting Scientist will be responsible for all aspects of the study contracting, designing, and monitoring; however, they may not have an in depth understanding of each individual process. This book aims at providing an overview of all aspects of study contracting and monitoring so that the Contracting Scientist clearly understands how all the pieces fit together, has an understanding of where studies can go wrong, and has a systematic template to follow to help guide in the conduct and oversight of a study to ensure they receive the highest quality study possible.

Unfortunately, contracting nonclinical studies is not taught in undergraduate or graduate school. Running studies in a non-GLP university setting is much different than running studies under a highly controlled and documented

---

1. Committee on the Judiciary United States Senate Ninety-Fourth Congress (January 20 and 22, 1976). Preclinical and clinical testing by the pharmaceutical industry, 1976. Examination of the process of drug testing and FDA's role in the regulation and conditions under which such testing is carried out: Part II. Joint hearings before the subcommittee on health of the committee on labor and public welfare and the subcommittee on administrative practice and procedure. US Government Printing Office.

GLP environment. There are some exceptions (e.g. universities that have set up separate GLP-compliant contract laboratories as offshoots of their normal operations); however, the vast majority of studies run by scientists during their undergraduate, graduate, or post-doctoral training are poorly controlled from a documentation perspective and the infrastructure to support GLP compliance is lacking. The studies are often state-of-the-art and push the cutting edge of science; however, they are often loosely designed and many changes are made to the studies either prior to study start or during the study. These changes are typically not documented in detail and may not be approved as required by the GLP regulations. In addition, the collection of study data is loosely controlled and there is often no quality control or quality assurance oversight. It is not uncommon for a procedure to be performed and an observation made and these are not documented until hours later, the next day, or even later, all of which would be major GLP deviations. As an example, a large animal study involved the administration of an infectious agent followed by administration of the test article. Since the CRO did not have the expertise to prepare the infectious agent, an expert professor from a highly respected university was enlisted to prepare the agent. The professor said that he had read the GLPs and "they didn't seem that complicated" and wanted to run his portion in compliance with GLPs. After a detailed discussion about the intricacies of GLP compliance, the professor was convinced to run his portion of the study as non-GLP. During the agent preparation, the professor started the timed preparation of the infectious agent. He neglected to record the start time and later back-entered the approximate time first on his glove and then transcribed this to his notebook. He threw out his glove and never signed or dated any entries. He believed he was conducting his portion "in the spirit of GLPs" but was not even close since the original data were on his glove, which he threw out, he approximated the start time after-the-fact, and he never signed or dated his entries.

Since scientists rarely come out of school with the knowledge of how to conduct GLP studies, this leaves the teaching process to individual companies. Depending on the company's resources and procedures, the teaching process may be in depth or it could be left up to the new Contracting Scientist to teach themselves. This book will help all levels of scientists understand what is involved with contracting and monitoring nonclinical studies and the critical areas that are most likely to result in poor studies. It may even provide ideas to seasoned scientists about ways to improve their current processes. This book covers all major aspects of a study from inception until completion and reporting, which will help the reader understand all the various processes, what can go wrong, how mistakes can be recovered from, and what factors play the most critical roles for a successful study. The authors have been involved with thousands of studies ranging from *in vitro* studies, acute and chronic rodent studies, to chronic large animal studies. Many of the studies followed routine guidelines and were "standard" CRO offerings; whereas, others were highly customized study designs that pushed the technical limits of the CROs. Regardless of the type of study, problems will arise in almost any study due to the numerous study-related functions that need to be conducted and recorded. Some problems are due to unforeseen circumstances that could not have been planned (e.g. the veterinary ophthalmologist is sick and cannot conduct protocol-driven eye exams). The majority, however, are due to technical conduct issues. This is not surprising, since people are ultimately conducting the study and are not infallible. Therefore, regardless of how well a study is planned, the success of a study often hinges on how well everyone involved with the study handles unforeseen circumstances and errors. When problems arise, it is best to keep a clear mind and think through all the possible options and consider how to document them so that an independent reviewer can reconstruct the event. Most problems are not terminal so long as an appropriate course of action is charted and followed.

GLPs AND NONCLINICAL STUDIES

Detailed information on Good Laboratory Practices (GLPs), how they impact the conduct and reporting of a study, and key differences in various GLPs (e.g. US Food and Drug Administration [FDA], Organisation for Economic Cooperation and Development [OECD]) are provided in Chapter 2. This book provides a general overview of GLP compliance, particularly issues that affect the scientific integrity or regulatory acceptance of a study; however, the hard compliance issues are left to other textbooks, guidance documents, and quality assurance (QA) personnel. The reader will be directed to appropriate references when necessary. Before delving into the specifics of GLPs, a few general points need to be raised about GLPs and the conduct of nonclinical studies. An important point that the reader needs to keep in mind is that GLP compliance does not guarantee a sound scientific study or regulatory acceptance. This book focuses first and foremost on running a sound scientific study that also complies with GLPs, when necessary. Many of the concepts of the GLP regulations are common sense and can improve the quality of any study, but claiming full GLP compliance when it is not can invalidate your study.

Depending on whom you talk to, the GLPs are a godsend, a curse, or a minor inconvenience. The GLPs were enacted due to rampant abuses at various laboratories running nonclinical studies in the 1970s. The abuses included the falsifying of data, poor animal housing conditions, and inadequate documentation of study functions, among many others. Unfortunately, all of these situations still occur and the last thing a Contracting Scientist wants is to be associated with any adverse situation that not only taints the study but the reputation of the Contracting Scientist and their company. In addition, GLP deviations may increase the cost of the study by either trying to "clean up" the study or repeating the study altogether. The GLPs aim at reassuring a regulatory authority (e.g. FDA, Environmental Protection Agency [EPA], European Medicines Agency [EMA]) that what is reported in a study report and recorded in the raw data truly represents what was done during the study. For the purposes of this book, a regulatory authority is an agency that reviews information on a product, such as a new drug, and determines if the information supports the marketing of the product. For example, the US FDA reviews manufacturing, efficacy, and safety (toxicology) data and studies to determine if an investigational new drug can be approved and sold for use as a new prescription drug.

Contrary to what some people think, GLP compliance is not a gold stamp that guarantees a successful study and all of the above abuses can still occur in a study that claims to be GLP compliant. GLP-compliant studies can fall way short of meeting regulatory acceptance from a scientific perspective and less scrupulous actions can be documented in what appears to be a GLP-compliant manner even though they are not accurately capturing what occurred. GLP compliance helps ensure what was done is recorded and reported accurately, but what was done may have been of very poor quality or certain critical study functions were never done since they were not included in the protocol, rendering the study unacceptable. In that case, you will have a report that accurately states what was done but no matter how good the reporting is, GLP compliance cannot fix errors in study design or conduct. Regardless of your view of GLPs, they have helped to improve the quality of nonclinical studies; however, focusing solely on GLP compliance minutia and missing the big picture of running a sound scientific study can lead to study failure.

Early in a drug development program, you may want to conduct studies that do not need to comply with the GLPs. Even later in development, some studies with unique designs, or portions of a given study, may not be conducted under GLPs. When it comes to defining a study in terms of GLPs, it is really black and white and either the study is conducted in compliance with the GLPs or it is not. Some laboratories or Sponsors may use the terms "GLP-like" or "In-the-Spirit of GLPs" when referring to non-GLP studies. These terms should be avoided and the study simply referred to as not being conducted in accordance with GLPs. From a scientific perspective, there is nothing wrong with running a non-GLP study so long as the laboratory maintains high quality procedures and documentation, as they would with a GLP study. The main difference is that the protocol, procedures, data, and report will not be formally reviewed by the Quality Assurance Unit (QAU). The key for running a non-GLP study is that you need to ensure that the laboratory adheres to its normal quality procedures and delivers a high quality study.

One last general point about GLP compliance is that you will encounter a wide range of Quality Assurance personalities and interpretations of the regulations at various laboratories and even within a single laboratory. The Quality Assurance Unit helps ensure that the study is conducted and reported according to the approved protocol, relevant GLPs, and laboratory Standard Operating Procedures (SOPs). The QAU reports to the facility management and therefore has independence from the personnel running the study. Theoretically, this is a beneficial setup since the QAU can make independent conclusions and report their findings to management without influence from the personnel running the study. However, the real world implementation of a QAU can vary widely. Some QAUs are very cooperative and work in a manner not only to help ensure GLP compliance but also improve the overall scientific integrity of the study. These QAUs are a pleasure to work with since not only do they provide open and honest feedback about potential compliance issues, but they are also willing to listen to all concerns and try to figure out procedures and processes that remain GLP compliant and also meet the scientific requirements of the study. Unfortunately, some QAUs act in a dictatorial manner and their word is the final word. They are unwilling to compromise and they see all compliance issues as black and white without providing any guidance on how a given procedure or process might be modified to maintain GLP compliance and still meet the scientific requirements of the study. These QAUs are very frustrating to work with and can lead to animosity among all parties involved with the study. It is important to keep in mind that most QAU auditors are not scientists and may not be experienced in the study design or all of the procedures in the protocol. Therefore, it is important for the QAU auditor to be able to work in a cooperative manner with the Sponsor and Study Director so that all scientific- and GLP-related issues are addressed appropriately. When auditing a new laboratory, a lot of effort needs to be made in assessing the working relationship of the QAU with the test facility management, scientific staff, and contracting company since this sole issue can lead to many obstacles in running a study that meets the scientific needs of the Contracting Scientist.

CROs AND NONCLINICAL STUDIES

CROs are in the business of making money. This is another factor to keep in mind when contracting studies since this can influence the level of oversight your study will receive. CROs make the most money by keeping animal rooms filled and using the least number of personnel to run the studies. This is less of a factor when running studies at internal company laboratories since the profit motive is not a major factor; however, the full utilization of resources can still play a role in justifying the existence of the laboratory. For CROs, they must balance profit motives with regulatory compliance and scientific integrity. This is not to say that a CRO will do the bare minimum in compliance and scientific integrity to maximize profit; however, there is a constant balance of these factors and, depending on the CRO, they may tip the scales one way or the other. Regardless, the Contracting Scientist must understand these competing forces and realize that simply signing a contract with a CRO to run a study on the CRO's own terms is a recipe for disaster. The Contracting Scientist must play an active role in every study they run and ensure the CRO is meeting all of their expectations. The advice in this book will help guide the Contracting Scientist in maintaining appropriate involvement and oversight of their study and ensure that the CRO fulfills all of their obligations and delivers a high quality study.

In order to maximize profit, a typical CRO will have a single person, regardless of their role, involved with many different studies. This ensures that everyone involved with the studies has minimal down time, thereby, maximizing personnel usage. The problem with this setup is that study personnel are often spread very thin and may have a hard time staying on top of all the various study functions for all the studies they are assigned to. This is where Contracting Scientist involvement with the study is essential. Even if the Contracting Scientist only interacts with the Study Director, the person at the CRO ultimately responsible for the successful conduct of the study, the Study Director will convey the needs of the Contracting Scientist to all study personnel and everyone is more likely to pay close attention to their study functions. If the Contracting Scientist does not play an active role, then people involved with the study are more likely to pay close attention to those studies where they know there will be routine and close oversight of their handling of the study. Therefore, it is best if the Contracting Scientist not only interacts routinely with the Study Director, but also makes visits to the laboratory and interacts with the various personnel involved with the study.

Employee turnover and covering weekend functions can be a major problem at CROs. Some CROs have excellent retention of employees whereas others have high turnover with "senior" study technicians having only two to three years of hands-on experience. In these high turnover CROs, the Contracting Scientist needs to be aware of the relative inexperience of most of the study personnel and how this can impact study conduct. When qualifying a new laboratory, a thorough review of the employees' training records and curricula vitae should be conducted to assess the experience of the people that will be involved with the study. Although experience *per se* is no guarantee for success, study personnel with experience are more likely to understand what can and cannot be done on a study, how to perform adequately both routine and complex study functions, and how to handle adequately problems when they arise. All of these factors play pivotal roles in the successful conduct of a study. When selecting a new laboratory or working with a new set of laboratory staff, the Contracting Scientist must carefully determine the experience level of all the people that will be involved with the study. In addition, the Contracting Scientist needs to be aware that studies that run over weekends are likely to use a different set of technicians than during the week. It is essential that anyone working on a study is familiar with the study functions and has been adequately trained. The authors have run many studies where weekend staff show up Saturday morning without ever reading the study protocol and fully understanding what needs to be done. Study personnel must be fully informed of the study requirements before the weekend so that they understand what needs to be done and how to do it. Since overall staffing on the weekends is typically reduced, the personnel working on the weekends should also be highly competent to conduct the study functions with minimal oversight or highly experienced personnel or supervisors should be present to help if any complications arise.

STUDY DIRECTORS

The Study Director is the person ultimately responsible for the conduct of the study and maintaining GLP compliance. He/she is also the Contracting Scientist's main contact at the laboratory. More details about the specific roles and responsibilities of the Study Director will be provided in subsequent chapters. The purpose of introducing the Study Director here is that the Contracting Scientist must establish a solid working relationship with the Study Director and clearly set expectations for not only their interaction but how the study should be run. As with CROs, some Study Directors excel at certain study functions but may fall short on others. Some Study Directors work their way up from study technician into the Study Director role. These Study Directors often excel with the logistics of running a study since they have hands-on experience of running many study functions. However, they may not have the scientific expertise for interpreting the findings. In contrast, the Study Director position may be the first job for a new PhD graduate. This person may have the scientific background and training for interpreting the study findings and writing a solid report; however, they may fall short on identifying the logistical hurdles with a study since they have never conducted hands-on functions in the laboratory, particularly under a GLP environment. Many PhD graduates have worked in a laboratory; however, fewer PhD graduates have hands-on animal experience due to the pressure to reduce animal experimentation in academic research. Even if a PhD graduate has animal research experience, it is unlikely that the research was conducted under rigorous GLP conditions. Therefore, the recent PhD graduate in a new Study Director role is unlikely to understand all the logistical issues that GLP compliance entails. Regardless of the background of the Study Director, it is important to understand their strengths and weaknesses so that an appropriate working relationship and expectations can be established.

As mentioned above, CROs maximize profit by having study personnel work on multiple studies. This also applies to Study Directors where some CROs may have a single Study Director conducting as many as 20 studies at a time. Even for relatively simple studies, this is a huge burden and the Study Director's time gets spread very thin. Some talented Study Directors may be able to handle this workload and truly stay on top of their studies; however, more often, Study Directors will pay more attention to those studies where they know the Contracting Scientist plays an active role. The Study Director will pay much more attention to a study where they know the Contracting Scientist will be making monitoring visits and requests routine study updates. Most Study Directors want to do the right thing and keep thorough track of all of their studies; however, the reality is that many of them are stretched thin and it is up to the Contracting Scientist to ensure that they receive the service and study oversight they need to ensure a sound scientific study. The best way to guarantee appropriate Study Director oversight of your study is to play an active role in the study so that the Study Director clearly understands your expectations. In addition, you need to ensure that the Study Director fosters an open line of communication between the QAU, Sponsor, and Study Personnel since this is critical for a successful study.

EXAMPLES OF STUDY ISSUES

In order to further strengthen the points that studies conducted at CROs require the active involvement of the Contracting Scientist and that GLP compliance does not guarantee a sound nonclinical study, various examples of problems encountered during the contracting and conduct of studies are summarized in the following sections. Although GLP compliance is not a guarantee of a successful study, it may help deter problems by requiring key protocol elements (e.g. proper planning of methods) and review and approval of the study protocol by all relevant parties. The following hypothetical examples come from the shared experience of the various authors and colleagues. These examples are not all inclusive and are simply meant to illustrate how studies can go wrong without appropriate Contracting Scientist oversight. Issues will still arise with appropriate study oversight; however, hopefully they can be caught earlier so that appropriate adjustments can be made to minimize the impact. Many of the issues outlined below would also be GLP compliance problems; however, some of them would have a profound impact on the scientific integrity of the study without having GLP compliance implications.

General Study Issues

In a GLP rodent carcinogenicity study, a very senior Study Director was selected to run the study. These studies involve daily exposure of the animals to the test article for up to 24 months. This particular study used daily oral gavage dosing, which is very labor intensive when compared to administration via feed. Other routine assessments typical for a carcinogenicity rat study were included such as body weight, feed consumption, and clinical observations. During an audit of this study, it was found that over the first 6 months, the Study Director never went back to the animal room to monitor the study and never reviewed the raw data. This particular Study Director was originally selected because of their scientific credentials and experience; however, it was clear that the Study Director was simply a figurehead and was not involved at all with the conduct of the study. There were systematic study documentation errors that should have been caught early in the study and corrected if the Study Director had maintained proper oversight of the study.

In contrast to the Study Director above, a Study Director for another study was selected for their hands-on experience with running complex studies, in this case a 6-month large animal study. On previous studies, the Study Director proved that he could coordinate non-routine study functions and successfully collect the data. Given their excellent ability to run the study and generate high quality data, expectations for a quality report were high. Unfortunately, it was a struggle getting a draft report from the Study Director and the original deadline passed. No time penalty was put in the contract so there was little incentive for the laboratory to meet a hard deadline. Once a final report was received, it was of extremely low quality with much missing information. It was clear that the Study Director was out of their element with scientific writing. The Contracting Scientist ended up essentially writing the report for the Study Director.

It was touched on above, but is worth mentioning again. Technician and Study Director turnover can be a huge detriment to a study, especially for chronic studies. A GLP large animal toxicity study that involved 12 months of oral dosing was being run at a laboratory with a historically high technician turnover rate. The rate was so high that a senior technician had just 2–3 years of experience. Several technicians conducting study functions had less than one year of experience. Although all technicians were trained on the study procedures they were conducting, there is a difference between conducting a study procedure you were just trained on and conducting a procedure you have done routinely for years. During the 12-month study period, the lead technician resigned and the Study Director was promoted to a different position within the laboratory. There were numerous dosing errors during the study and some study functions were mistakenly skipped. All of this was attributable to poor oversight of the study by the Study Director, particularly during the transition period to the new Study Director, and lack of coordination of study activities by the lead technician after their resignation.

Protocol

The development of a protocol that meets scientific, study logistic, GLP, other regulatory, and animal welfare requirements can be very difficult, especially for complex studies that do not fit traditional "guideline" formats. It is very important to take sufficient time to draft the protocol so that all personnel, both at the testing facility and at the Contracting Scientist's company, are allowed time to review the protocol and submit comments to the Study Director. Sometimes protocols can be written that sound great to the Study Director and Contracting Scientist but may not be logistically possible. The best protocols are those that are reviewed and re-reviewed by everyone involved prior to them being issued. This also cuts down on the number of protocol amendments that may be necessary.

A 3-month large animal study was being conducted in a country that has very strict animal welfare requirements, in particular, animals have to be group housed at all times. This caused major problems since the study was being used to support a US regulatory submission and the US regulatory body requested single housing of the animals so that individual feed and water consumption and clinical observations could be recorded. A one-time exemption to the group housing rule was granted but this caused a 4-month delay in the project. This protocol also suffered from the lack of adequate personnel at the laboratory.

For animal observations conducted on the weekend, the laboratory was going to use security personnel to conduct the observations even though they were not trained for this! This compliance issue was not obvious from the writing in the protocol and only became apparent during teleconferences to discuss the protocol, study expectations, and roles and responsibilities. An interesting aside to this study is that this was a "GLP-certified" laboratory. GLP compliance is different in the USA versus other countries since laboratories in the USA are not certified whereas, laboratories in other countries typically undergo a GLP certification process, after which, it is assumed that any GLP study coming out of the laboratory is GLP compliant. This assumption was obviously not correct for this particular study.

Given the cost and time of running nonclinical studies, there is always the temptation to add endpoints to a single study to meet the scientific requirements instead of conducting separate studies that assess the endpoints independently. The benefit of the single study approach is that all the endpoints are assessed in the same set of animals, providing for more powerful correlative analyses. However, adding a lot of extra endpoints can present major logistical issues, especially under GLPs where all study functions have to be documented in detail. In a GLP 13-week subchronic toxicology large animal study, all of the standard guideline endpoints were included, including a complete necropsy. In addition, several interim bleeds, interim sperm collection, and many additional terminal sample collections during necropsy were included in the protocol. The interim bleeds pushed the blood volume withdrawal limits and some of the animals had anemia. The interim sperm collections were labor intensive and interfered with the timing of other study functions. The necropsy ended up being a major failure since too many collections and divisions of samples occurred in a limited timeframe. Not only did documentation errors prevail, but sample integrity was compromised since some tissues and samples were not processed in a timely manner. In hindsight, it would have been best to conduct two separate studies or at least use satellite animals that did not have to undergo all of the routine study procedures.

Some issues can arise that no matter how much you plan for them in the design of the protocol, they still rear their ugly head. A GLP subchronic large animal study involved topical administration of the test article for 6 months. The formulated test article had low viscosity and spread very easily. There was a concern that during the handling of the animals for the dosing procedures and clinical observations, some test article might transfer to the technician's gloves and be subsequently transferred to the next animal. Therefore, the protocol specified that technicians change gloves between every animal. Unfortunately, pharmacokinetic analyses of the blood showed that control animals had been exposed to the test article. It was determined that the low level exposure was either due to animal care technicians moving animals during cage changes since they were not required to change gloves or some of the test article volatilized off the animals and exposed the control animals via inhalation. The protocol should have included the requirement of *any* person contacting an animal to change gloves between animals; however, it was believed that the test article would not transfer off the animal several hours after application, which is when the animal care technicians would make cage changes.

Test Article

Test article sourcing, characterization, and formulation are common areas for problems. These are basic necessities of a well-run study and the success of all of the subsequent study functions hinge on using the appropriate test article. Depending on the stage of product development and the study requirements, the test article may be a single, poorly characterized active ingredient or a well characterized final formulation. It is important to understand the study require- ments so that you use test articles that are sufficiently characterized for the study purposes and meet the GLP requirements. In a large animal study, the final proposed marketed formulation was used for dosing the animals. The time- lines for the project were very tight and initial stability results for the proposed marketed formulation showed borderline stability. The decision was made to run the study with a modified formulation without having upfront stability informa- tion. Several months into the study, the test article was tested for stability and failed and the study had to be repeated. Since the success of any study hinges on the test article, it is advisable to insist on having the test article and formulations characterized up front to the level required by the study.

Methylcellulose and other cellulose derivatives are often used to thicken oral test article preparations to aid stability and homogeneity. This is essential when a test article forms a suspension instead of a solution, which often occurs for low solubility test articles administered at high dose levels. If a suspension aid was not used, the test article would quickly settle out, even with constant stir- ring, making it nearly impossible to draw up homogeneous dose formulations. In a GLP 13-week rodent study, the test article was formulated with methylcel- lulose of a medium viscosity. Methylcellulose is available in a variety of differ- ent viscosities to suit different needs. Unfortunately, the formulation technician selected a methylcellulose of a low viscosity for the daily dose preparation and this was not discovered until one month into the study. All of the GLP test article formulation work was done with the medium viscosity preparation and could not be applied to the low viscosity preparation. Separate stability and homoge- neity studies were performed on the low viscosity preparation but the results did not pass the study requirements and the study had to be repeated.

An ocular tolerance study was being conducted in a non-rodent species and involved injection of sterile test article into one eye of each animal. The protocol specified that the lyophilized test article had to be rehydrated under sterile conditions using sterile solutions. The protocol did not clearly state the actual procedure and the formulations department prepared the solution using sterilized saline but did not sterile filter the final solution. After adverse effects were noted in some injected eyes in both the control and treated groups, it was concluded that bacterial contamination occurred during the preparation since the lyophilized test article was not sterilized. It was concluded that sterile filtration of the final solution would have solved this problem and should have been clearly specified in the protocol.

In-Life

In a typical GLP animal study, the vast majority of study functions and documentation occur during the in-life phase, which includes the acclimatization, dosing, and observation phases until the start of the necropsies, when applicable. It follows then that the greatest potential for mistakes occurs during this labor-intensive phase. It is almost inevitable that some mistakes will be made during the thousands of individual study functions and observations that need to be made during a typical study. Computer systems can help decrease errors but errors can still occur since not every function is controlled by the computer (e.g. the computer does not conduct oral gavage dosing and the computer does not handle the animals for observations). In addition, some functions are not amenable to computer tracking or control. Some of the errors that occur during the in-life phase may be relatively simple errors that do not impact the integrity of the study; whereas, others may compromise the whole study.

A relatively simple, but potentially study ending mistake is not verifying that the right animal is being dosed and observed. The use of implantable radio frequency identification (RFID) tags decreases the potential for animal identification mistakes since the animal must be "scanned" to confirm the correct animal is being dosed and observed. If the wrong animal is selected, the computer system will warn you and not allow you to dose or observe the animal. Therefore, RFID tags should be used whenever a laboratory has this functionality. However, some laboratories still use manual means of animal identification. In a GLP subchronic non-rodent study, animal technicians were supposed to verify the animal ear tag number with the cage card and call out the animal number to the second technician conducting the dosing and making clinical observations so that they could verify the right animal was being used. Unfortunately, these procedures were not followed and even though the right cage was being selected, for some reason the wrong animal was in the cage, probably due to a mix up during cage changes. Since the animal ID was not being verified, several animals received the incorrect dosage for over a week and the study had to be repeated.

In a subchronic large animal study, the Study Director decided to add an additional clinical observation between the protocol-specified a.m. and p.m. observations after the study started. The Study Director never notified the Sponsor about this addition and never made a protocol amendment to add the observations. Although the Study Director was well intentioned, this was a GLP deviation and could have placed undue stress on the animals due to the additional handling. The test article being tested caused neurological stimulation and the effects were aggravated by the additional animal handling, which is why the protocol listed only two clinical observations.

As mentioned above, computer guidance and tracking of study functions helps to reduce errors but, unfortunately, there are still problems that can arise, especially if technicians are not paying attention to computer messages and are blindly entering information. In a GLP large animal study, all study functions appeared to be proceeding according to the protocol with no deviations recorded to date. Protocol and study deviations will be discussed in detail in Chapter 10; however, a brief description deserves mention at this point. Essentially, deviations are study functions that did not meet the protocol-specified requirements (e.g. the blood collection time windows were exceeded) or were not specified in the protocol. Deviations may or may not have a detrimental impact on the study. In this particular study, during a monitoring visit and review of the raw data, it was noted that one high dose animal died on study. The first problem was that the Contracting Scientist was never notified of this dead animal even though immediate notification of serious adverse events was clearly specified in the study contract. The second problem is that this animal continued to be dosed, at least according to the computer records, for five days after it died! It ends up that the technician recording dosing and clinical observations just hit enter repeatedly without paying attention to animal numbers or if animals were actually dosed. This seems almost impossible with a computer system that requires multiple inputs and confirmations; however, the technician basically bypassed the safeguards by rushing through the procedures and not paying attention to what was being asked by the computer system.

A 13-week rodent study was being conducted at a laboratory that had state-of-the-art computer control of study functions including RFID tags for animals and bar coding for all dose formulations. Since the computer required the verification of animals and doses, it appeared that everything was running flawlessly. When the first pharmacokinetic sampling results came back, the low and high dose groups appeared to be reversed. Dose formulation samples were collected during this same week and were sent for analysis. It ends up that even though bar coding was being used, the formulations department mislabeled the dosing vials for this week resulting in incorrect dosing of the low and high dose groups.

Proper adherence to the protocol, amendments, and laboratory SOPs are common problems, especially when technicians new to the study are assigned to conduct study functions (e.g. weekend technicians that have not been assigned to the study previously). It is imperative that all people involved with the study understand the protocol and all amendments. In a GLP 28-day rodent study, an amendment to the protocol specified that clinical observations were to be made four times a day at 1, 2, 4, and 8 hours after the dose. The amendment was made since adverse clinical signs were occurring and had to be captured more routinely to see if dose adjustment was required. A weekend technician new to the study and their supervisor failed to read the amendment and only conducted two clinical observations in the a.m. and p.m., which were the timeframes specified in the original protocol. This was a clear deviation but also had the impact of not being able to assess accurately if the high dose should be lowered in a timely manner to prevent animal suffering.

The reporting of deviations within a laboratory (e.g. from the technical staff to the Study Director) and to the Contracting Scientist can cause problems since if they are not reported in a timely manner, procedures may not be able to be adjusted in time. Some deviations do not require modification of the study since they are relatively minor and occur infrequently. However, a deviation may require an amendment to the protocol in order to prevent future deviations. In a GLP 13-week large animal study, pharmacokinetic blood sampling was required in a large number of animals at tight timeframes. In past studies, the frequent blood sampling was successfully conducted; however, this study utilized a very viscous oral dose solution that was difficult to pull up into the syringe and administer. Therefore, even though staggered dosing was used to meet the pharmacokinetic sampling requirements, many doses were administered off of their specified timeframe and this compounded into missing the sampling windows for the pharmacokinetic blood draws. These deviations were not reported to the Study Director for over a month. By this time, a second round of pharmacokinetic sampling was conducted which experienced the same sampling issues. If these deviations were reported in a timely manner, the Study Director could have notified the Contracting Scientist and made adjustments to the blood sampling schedule to avoid this second round of deviations.

Unexpected issues can also come up for other regulatory compliance reasons besides GLPs. An animal welfare issue came up in a 6-month large animal study being conducted with a topical product. The product was being applied to a defined area on the back of the animals. The protocol specified that animals had to be housed in runs with plastic barriers attached to the sides of the runs. The plastic barriers prevented the animals from scratching their backs on the metal fencing that divided each run and also prevented animal-to-animal contact to avoid transfer of test article between animals. Unfortunately, the animals found a way to rout with their nose under the plastic barriers. This natural behavior caused them to scrape their nose on the edge of the plastic and get small scratches. In addition, the animals scratched the treated area of their backs on the water supply lines protruding from the wall. The USDA inspected this study and considered the scratches a major animal welfare issue even though the scratches were superficial. The runs were further modified by lowering and smoothing the edges of the plastic barriers and raising the water supply lines above the height of the animal's backs. Even with these modifications, the animals still managed to scratch their noses and backs, albeit at reduced rates.

Necropsy

If your study involves necropsies, appropriate planning is required, especially if you have non-routine samples that you will be collecting. Necropsies typically occur at the end of the study, unless there are interim sacrifices, and can involve a lot of collections and observations over a short time frame. Appropriate planning, pre-necropsy meetings, and even practice necropsies for complicated studies, can decrease the number of errors during this critical phase.

In a GLP chronic rodent study, terminal necropsies were being conducted that involved routine gross necropsy assessments, organ weights, and tissue collections for histopathology. In addition, specific tissue samples were being collected for mRNA genomics analyses using microarrays. Since timely sample collection and flash freezing in liquid nitrogen are critical for preserving mRNA integrity, these collections had to be integrated into the normal pathology procedures. A test necropsy was run on two animals prior to the actual necropsies to ensure that the procedures would meet the timed necropsy windows. During the test necropsies, we found that the genomics sample collection and processing took more time than expected and would have resulted in numerous deviations if we did not amend our dosing timelines in order to meet the timed necropsy windows.

In a GLP non-rodent ocular study, the left eye was being injected with the test article and the right eye served as an untreated control. The protocol did not specify how the eyes should be collected during the necropsy although it was implied that some method should have been used to distinguish the left treated eye from the right untreated eye. The eyes were collected and placed in the same fixation jar without identifying left from right. Although treatment related effects appeared to be observed, these could not be definitively attributed to the test article since the eyes could not be distinguished from each other and the study had to be repeated.

Reporting

As mentioned previously, some laboratories might conduct a solid study; however, their reporting leaves a lot to be desired. In a relatively complex GLP subchronic large animal study, a draft report was required by a fixed date. The study went smoothly and interim monitoring visits did not reveal any issues with the data. The draft report was received by the specified date but was poorly written, was outright missing sections of information, and some contributing scientist reports for various study functions were also missing. This particular Study Director came up through the technical ranks and was excellent at running the study but obviously report writing was not one of their strengths. The Contracting Scientist ended up rewriting the draft report for the Study Director in order to meet the timelines.

In some nonclinical studies, regulatory authorities may require that copies of hand captured raw data be submitted as an appendix to the study report as part of the new drug approval process. This is not a major problem for electronically-captured data since the regulatory authorities just require the typical individual animal data tables for the various endpoints (e.g. clinical observations, body weight, feed consumption, clinical pathology). A new laboratory was selected to run one of these unique studies. Even though the protocol clearly stated this unique reporting requirement, they failed to include copies of the raw data in the draft report. Since this particular laboratory captured most of their data by hand, it was a major undertaking to compile all of the raw data for this 1-year large animal study and caused a delay in the issuance of the final report.

In a GLP 13-week rodent study, we had to assess the blood level of a hormone that was potentially affected by the test article. The laboratory repeatedly told us that the assay would be validated under GLPs prior to the start of the study. When we received the contributing scientist report, the GLP compliance statement indicated compliance with the regulations; however, the study timeline showed that the method validation occurred after the study samples were analyzed. Therefore, the validity of the results was unknown.

CONCLUSION

This chapter presented an overview of why appropriate Contracting Scientist oversight and communication between the laboratory, Study Director, and other relevant personnel is essential to the success of a study. Simply signing a contract with a laboratory to run a study and then not being involved with the conduct of the study is a recipe for disaster, even for relatively simple, short-term studies. When the Contracting Scientist plays an active role, not only do they send a clear message to the laboratory that they want a high quality study and will hold the laboratory accountable, but they can catch mistakes early so that appropriate corrective actions can be made. The following chapters cover the various critical areas of running and reporting a study so that the Contracting Scientist has the knowledge and can devise the best plan of action for running and monitoring their study.

Good Laboratory Practices

Joe M. Fowler BS, RQAP-GLP*, William F. Salminen PhD, DABT, PMP†
and James Greenhaw BS, LAT*

*National Center for Toxicological Research, FDA, Jefferson, AR, †PAREXEL International, Sarasota, FL

> ## Key Points
> - The GLPs help ensure that the study is conducted and data are collected and reported in compliance with the study protocol
> - GLP compliance does not guarantee a sound scientific study or regulatory acceptance
> - The GLPs aim at reassuring a regulatory authority that what is reported in a study report and recorded in the raw data truly represents what was done during the study
> - The Quality Assurance Unit (QAU) helps ensure that the study is conducted and reported according to the approved protocol and relevant GLPs and laboratory Standard Operating Procedures (SOPs)

The Good Laboratory Practices (GLPs) were enacted to help reassure regulatory authorities that the nonclinical study reports being submitted to support registration of the test article accurately represented what was actually done during the study and that the results were accurate. Prior to the enactment of the Food and Drug Administration (FDA) GLPs in 1976, the assumption at the FDA was that study reports submitted by Sponsors to support a new product application (e.g. new drug) accurately described the study conduct and precisely reported the study data. Unfortunately, during "for cause" inspections of studies conducted at a variety of laboratories, many data inconsistencies and evidence of unacceptable laboratory practices were discovered. Major defects in the design, conduct, and reporting of the studies were found. Since these studies played a pivotal role in FDA's assessment of the safety of new drug products, doubts were raised about the validity of past and ongoing safety assessments that incorporated results from these flawed studies. The GLPs were enacted as a way and means of ensuring the validity and reliability of nonclinical studies submitted to the FDA.

Nonclinical Study Contracting and Monitoring. http://dx.doi.org/10.1016/B978-0-12-397829-5.00002-8
2013, Published by Elsevier Inc.

GLP compliance assures a regulatory reviewer that what they are reading and reviewing actually represents what was done and that the results are accurate. An easy way to think about the role of GLPs from a reviewer's perspective is encompassed by the statement, "If it is not documented, it did not happen". GLPs help the reviewer understand and recreate all of the functions and data that were conducted and collected during the study. From the reviewer's perspective, even if the protocol stated a given function was supposed to be conducted, unless it is accurately recorded and documented, they have no way of knowing if it was done or done in compliance with the protocol. It is important to keep these points in mind since some Quality Assurance Units (QAUs) will focus on minutia and lose the big picture of the ultimate purpose of the GLPs. A QAU reviewer might raise a huge issue with a procedure that slightly deviated from a Standard Operating Procedure (SOP) but may have no issue with important descriptions of the conduct of certain study functions simply because they were not listed in the protocol or SOPs (e.g. no description is provided for the difficult drawing up and oral gavage dosing of a viscous test article that may impact the amount of test article administered to certain animals). If the Study Director or Contracting Scientist feels that certain study conduct details are important, then those details should be listed in the study protocol or in an appendix to the study protocol since this will allow the QAU to understand clearly the critical aspects of the study that need to be carefully monitored.

The most common GLPs used for nonclinical studies are those enacted by the US FDA (21 Code of Federal Regulations part 58) and the Organisation for Economic Cooperation and Development (OECD). They have similar structures and requirements; however there are important distinctions, such as coverage of multisite studies in the OECD GLPs but not by the US FDA GLPs. Copies of the various GLPs can be found online at:

- US FDA 21 CFR Part 58
  (http://www.accessdata.fda.gov/scripts/cdrh/cfdocs/cfcfr/cfrsearch.cfm?cfrpart=58)
- OECD
  (http://www.oecd.org/document/63/0,3746,en_2649_34377_2346175_1_1_1_1,00.html)
- EU
  (http://ec.europa.eu/enterprise/sectors/chemicals/documents/classification/laboratory-practice/)
- Japan
  Japanese Good Laboratory Practice Standards for Safety Studies on Drugs (Ordinance Number 21 of the Pharmaceutical Affairs Bureau, Ministry of Health, Labor and Welfare, Japan; effective April 1, 1997)

When selecting a new Contract Research Organization (CRO), it is important to determine their GLP compliance strategies along with their past performance. Depending on the size of your organization, you may have a Quality Assurance (QA) group that can visit the laboratory and conduct a comprehensive audit of the CRO's GLP procedures. However, if GLP compliance performance is left up to the Contracting Scientist, the information provided in this chapter will be helpful in assessing a CRO's GLP compliance procedures.

The GLPs are broken down into various sections and each one will be reviewed below. Key differences between the US FDA and OECD GLPs will be highlighted in the various sections. At the end of the chapter, a monitoring checklist is provided to help the Contracting Scientist determine the level of GLP compliance.

The following overview of the GLP sections is provided in the most common sense terms possible. This is meant to provide the Contracting Scientist with a foundation for understanding the GLPs and is not meant as an academic debate about the specific meaning and interpretation of the more controversial aspects – these are left to other forums about the interpretation and implementation of GLPs. When reading the following information, the reader should have copies of the relevant GLPs for reference since the information provided below is not meant as a GLP reference but as an overview of the key aspects of each section and a comparison between the US FDA and OECD GLPs. The Contracting Scientist should also understand that there are industry standards, guidelines, and submission requirements that may not necessarily be required by the GLP regulations themselves. Often times during the "GLP qualifying inspection", these industry standard interpretations of the GLPs surface. It is important that the test facility management and QAU evaluate the regulatory risks involved with the implementation of the different compliance strategies.

Although the GLPs can seem relatively simple, their actual implementation is very complex and should not be taken lightly. The actual logistics of implementing all the requirements specified in the GLPs and ensuring that everyone is trained and complies with the requirements is a major undertaking that requires constant vigilance. CROs spend a lot of time and resources implementing and complying with GLPs and they spend a substantial portion of every study on GLP compliance. If GLP compliance was easy, many more laboratories would be conducting GLP-compliant nonclinical studies since not only are they profitable but they help bolster the reputation of a laboratory.

An important aspect that the Contracting Scientist should be aware of is that some laboratories may say they conduct studies according to "good practices", "in the spirit of GLPs", or other similar terminology. It is important to differentiate clearly between formal, regulatory use of the term GLP and the general application of "good practices" or similar terminology in scientific investigation. It must be clearly understood that only adherence to and compliance with all the requirements of the GLPs constitutes real compliance with the GLPs. Therefore, the use of similar terminology to describe quality practices outside the scope of the GLPs should be avoided.[1]

---

1. World Health Organization (WHO), 2009. Handbook: Good Laboratory Practice (GLP) - Quality Practices for Regulated Non-clinical Research and Development. WHO Library: available at: http://www.who.int/tdr/publications/documents/glp-handbook.pdf.

Another important distinction is quality control (QC) versus quality assurance (QA). Every study, even non-GLP, must have QC procedures in place. QC entails reviewing the data routinely to ensure that it was captured correctly. Often times, QC is conducted by people who are intimately familiar with the procedures. For example, a study supervisor may conduct a daily QC of all of the data that their technicians collected during the day. In contrast, QA refers to the review of data and procedures by the QAU of the laboratory. This is a GLP requirement; whereas the QC review is not technically required by the GLPs. The QA reviews are typically conducted less frequently than the QC reviews and typically do not involve auditing all of the data. For example, some QAUs will conduct a thorough audit of 10% of the raw data unless they find errors and then they will continue to audit additional data. In addition, the QAU reviewer may not be familiar with the procedures used to collect the data. Any successful study must have a combination of robust QC and QA procedures in place to ensure that the data are captured and recorded accurately.

US FDA (21 CFR PART 58) AND OECD GLPs

The US FDA GLPs will be used as the general template for reviewing the various sections of the GLPs since many studies are conducted in compliance with US FDA GLPs to support drug registrations in the USA and elsewhere throughout the world. Highlights of differences from the OECD GLPs will be provided since the OECD GLPs use some different terminology and provide guidance on conducting multisite studies, which is lacking from the FDA GLPs. It is beyond the scope of this chapter to provide a detailed comparison between all the different GLPs throughout the world and this is left to other textbooks and guidance documents on GLPs. The focus of this chapter will be on the application of GLPs in whole animal studies; however, it is important to note that the GLPs are not restricted to just those types of studies. GLPs apply to both *in vivo* and *in vitro* studies and also to studies conducted with plants and microorganisms as well as animal, plant, and microorganism parts.

In many sections of the GLPs, fairly generic guidance or criteria are provided. It is ultimately left up to the laboratory to determine the best methods and procedures for many items and often there is no one simple answer. For example, one laboratory may choose to use radio frequency identification (RFID) chips for animal identification; whereas, another will use ear tags and both methods are GLP compliant. However, there are other aspects of the GLPs that are black and white (e.g. a minimum set of SOPs is required and test articles must not be stored in animal rooms). It is important to be able to understand which aspects have flexibility and which are fixed. In addition, the Contracting Scientist should understand that there are industry standards, guidelines, and submission requirements that may not necessarily be required by the GLP regulations themselves and can impact the design and conduct of the study.

A final note is that US regulation 21 CFR Part 11 covers electronic records and electronic signatures. This part applies to a wide array of FDA-related uses, including the US FDA GLPs. Essentially, if the requirements of this part are met, the FDA considers the electronic signatures equivalent to full hand-written signatures and allows electronic records to be used in lieu of paper records. These specific requirements are highlighted in the relevant sections. Even though they may not necessarily be listed in the US FDA GLPs, Part 11 applies since it is the overarching regulation whenever electronic records or signatures are used. In order to comply with Part 11, all predicate rules for data integrity required by 21 CFR Part 58 (GLP) should be met.

SUBPART A – GENERAL PROVISIONS

This subpart outlines various definitions and general responsibilities when conducting a GLP-compliant study.

Section 58.3 (Definitions)

This section defines various terms that are used throughout the GLPs. A few terms deserve special attention.

- *Nonclinical Laboratory Study*: the FDA GLPs specifically use this terminology; however, for the purposes of this book it is synonymous with "preclinical study". This is any *in vivo* or *in vitro* experiment conducted with the *Test Article* in the *Test System* under laboratory conditions. The term specifically excludes studies utilizing human subjects or clinical studies or field trials in animals. Importantly, basic exploratory studies and physical or chemical characterization studies are not covered.
- *Test Article*: this is the regulated product or product's active ingredient that will be tested in the study.
- *Test System*: this is the experimental model (e.g. whole animal) that is being used.
- *Control Article*: this is the article, other than the *Test Article*, that is administered to the *Test System* for the purpose of establishing a basis for comparison with the *Test Article*. For most animal studies, this will be the vehicle (e.g. water, saline, aqueous methylcellulose) used to formulate the *Test Article* so that it can be appropriately administered. Some studies may also use a positive control test article to ensure that the model is responding appropriately.
- *Sponsor*: the FDA GLPs specifically use this terminology; however, for the purposes of this book it is synonymous with *Contracting Scientist* and their company or organization. This is the *Person* who initiates and supports, by provision of financial or other resources, a *Nonclinical Laboratory Study*. It can also be the *Person* submitting the study to the US FDA in support of a new drug application or investigational new drug application if the study has already been completed.

- *Study Director*: this is the individual responsible for the overall conduct of a *Nonclinical Laboratory Study*. The *Study Director* is typically located at the laboratory running the study and coordinates all of the study functions. They are responsible for assuring that all applicable good laboratory practices are followed.
- *Person*: this term is confusing since it is not necessarily a single individual as it can be an individual, partnership, corporation, association, scientific or academic establishment, government agency, or any other legal entity.
- *Testing Facility*: the FDA GLPs specifically use this terminology; however, for the purposes of this book, it is synonymous with CRO or laboratory. The *Testing Facility* is the *Person* (as defined above) that conducts the *Nonclinical Laboratory Study* by using the *Test Article* in the *Test System* (e.g. conducts dosing of animals with the *Test Article*).
- *Quality Assurance Unit (QAU)*: this is the individual or group that is designated by *Testing Facility* management to perform quality assurance duties. They must not be the same person as the *Study Director* on a given study and must be completely independent from personnel conducting the study.
- *Raw Data*: many debates have occurred over the definition of raw data. It is essentially the medium where the original observation was recorded. This could be on paper, in an electronic medium (e.g. computer system), a photograph, on a person's glove, etc. It is wherever the observation was first recorded. It is important to note that if an original observation is changed, the original recording must not be obscured. The change must be justified with a reason for the change and the person making the change identified on the date of the change. With electronically recorded data, this is often referred to as an audit trail.
- *Study Initiation Date*: this is the date the protocol is signed by the *Study Director.*
- *Study Completion Date*: this is the date the final report is signed by the *Study Director.*

Differences from the OECD GLPs

The OECD GLPs use some different terminology.

- *Test Site*: this is any location where a given phase of a study is conducted. It is particularly important for multisite studies since the *Testing Facility* includes the site where the *Study Director* is located and all the *Test Sites* where different portions of the study are conducted.
- *Principal Investigator*: this term applies to multisite studies and is the person at a *Test Site* that acts on behalf of the *Study Director* for the given phase of the study. An important point to note is that the *Study Director* cannot delegate to the *Principal Investigator* the approval of the study plan and amendments, approval of the final report, and compliance with GLPs. However, the

Principal Investigator is responsible for complying with the GLPs for their assigned phase of the study.

- *Study Plan*: this term is used instead of protocol.
- *Experimental Starting and Completion Dates*: these are the dates on which the first study specific data are collected and the last date on which data are collected, respectively. These are in addition to the *Study Initiation* and *Study Completion Dates*.
- *Test and Reference Items*: these terms are used instead of *Test* and *Control Articles*, respectively.

Section 58.10 (Applicability to Studies Performed Under Grants and Contracts)

If you are sponsoring a GLP study and are utilizing a third party (e.g. CRO), you must tell the third party that the study must comply with GLPs.

Section 58.15 (Inspection of a Testing Facility)

A testing facility conducting a GLP study should allow the FDA to inspect the facility and all records and specimens that are required to be maintained with the study. The FDA can make copies of the records. An important exception is that the FDA may *not* inspect or copy QAU audit records but they can assess QAU procedures.

SUBPART B – ORGANIZATION AND PERSONNEL

This subpart outlines the personnel and facility organization required for conducting a GLP study and everyone's specific responsibilities. Figure 2.1 outlines the key personnel and their reporting lines for a GLP study.

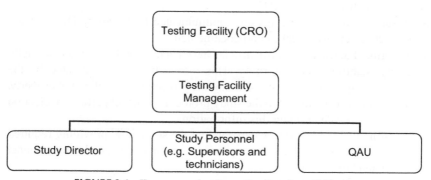

FIGURE 2.1   Key personnel and reporting lines for a GLP study

Section 58.29 (Personnel)

1. Each person involved with the study must have education, training, and experience that allows them adequately to perform their duties and a training file including their experience and job description must be maintained. CROs spend a lot of time ensuring their personnel are trained and that the training documentation is maintained.
2. Personnel must be trained in GLPs.
3. The laboratory must have a sufficient number of personnel, they must have appropriate sanitation procedures in place, and they must wear clothing appropriate for their duties. Any person who has an illness that may affect the study (e.g. contamination of test and control articles and test systems) must be excluded from study functions until the illness is resolved.

Computer Systems (21 CFR Part 11)

If the CRO uses computerized systems, the following aspects are important:

1. Personnel involved in the design, development, and validation of the computer system must be identified.
2. Personnel responsible for the operation of the computer system must be identified.
3. Personnel using the computer system must be trained.
4. Personnel entering and modifying data must be identified.

Section 58.31 (Testing Facility Management)

Testing Facility Management is responsible for the facility and ensuring resources are provided to implement the GLPs. The roles of Testing Facility Management include:

1. Designating a Study Director, controlling their workload, and replacing them promptly when necessary.
2. Assuring that there is a QAU.
3. Assuring that QAU findings are communicated to the Study Director and corrective actions are taken and documented.
4. Assuring that test and control articles have been tested for identity, strength, purity, stability, and uniformity. This may actually be conducted by the Sponsor, especially when considering the protection of intellectual property. This should be addressed in the study protocol so that all parties are clear on who will provide or retain this information.
5. Assuring that the laboratory can conduct the study (e.g. providing required personnel, resources, facilities, equipment, and materials) and that personnel understand their duties.
6. Providing GLP and technical training.

Differences from the OECD GLPs

The OECD GLPs have some additional responsibilities for multisite studies. For multisite studies, Testing Facility Management is the overarching management for the whole study and they must:

- Ensure that Principal Investigators are appropriately trained, qualified, and experienced to supervise the delegated phase of the study and if replacement is required, it should be documented.
- Ensure that clear lines of communication exist between the Study Director, Principal Investigator(s), QAU, and study personnel.

Section 58.33 (Study Director)

Each study must have an appropriately qualified Study Director assigned to it. An important point is that the Study Director is the single point of control for the study and must remain actively involved with the study. The Study Director must assure:

1. The protocol, and any change, is approved and followed.
2. All data are accurately recorded and verified and are collected according to the protocol and SOPs.
3. Unforeseen circumstances are noted and corrective actions taken and documented.
4. Study personnel are familiar with and adhere to the study protocol and SOPs.
5. GLPs are followed.
6. Study data, final reports, and specimens that are amenable to storage are transferred to the archives during or at the close of the study.

Differences from the OECD GLPs

For multisite studies, the Study Director is responsible for the overall study, including each phase conducted at a Test Site even if a Principal Investigator is used. It is important to understand this responsibility since it is ultimately the Study Director's responsibility to ensure a study's compliance with GLPs. Multisite studies are often complicated by different views at the various Test Sites of how to comply with different aspects of the GLPs but it is ultimately up to the Study Director to make the final decision on GLP compliance. When a Principal Investigator is used, the Principal Investigator must ensure that the delegated phase of the study is conducted in accordance with GLPs.

Section 58.35 (QAU)

The testing facility must have a QAU that monitors the study to assure management that the study is being conducted in compliance with the GLPs. The QAU must be independent from the personnel conducting the study (e.g. Study

Director and technicians). An important point to note is that for smaller laboratories, a Study Director for one study can actually serve as a QAU auditor on another study so long as for a given study, the QAU and Study Director functions are independent.

The QAU must have SOPs that cover the following:

1. Maintaining a copy of the master schedule sheet that lists all studies being conducted at the laboratory. The QAU does not have to generate the master schedule but they must maintain a copy. Also, there are conflicting opinions on whether or not the master schedule should contain non-GLP nonclinical studies. Since they are non-GLP, they do not necessarily have to be included; however, a significant portion of a Study Director's workload may be non-GLP studies. In order to assess adequately Study Director workload, these non-GLP studies should ideally be listed on the master schedule.

2. Maintaining copies of all GLP protocols and amendments.

3. Inspecting studies at appropriate intervals, maintain inspection records, and immediately notify the Study Director and management of issues along with providing periodic status reports.

4. Determining that no deviations from approved protocols or SOPs were made without authorization and documentation. The term "authorization" is confusing since many deviations are made by mistake without appropriate authorization. In these cases, the deviations must be acknowledged by the Study Director, their impact on the study assessed, and a plan of action to prevent similar deviations implemented.

5. Reviewing the final study report to ensure that it reflects what was done and the results match the raw data.

6. Preparing and signing a statement that is included in the final report that specifies the dates inspections were made and findings reported to management and the Study Director.

7. Inspecting computer operations.

The FDA can inspect QAU records but it is limited to inspection dates, study inspected, phase or segment of study inspected, and name of individual performing the inspection. The FDA can also inspect QAU written procedures and require that Testing Facility Management certify the procedures are being followed and inspections are being conducted. The FDA cannot inspect QAU audit findings.

Differences from the OECD GLPs

For multisite studies, the QAU at the Testing Facility is responsible for the overall quality assurance of the study (i.e. Lead QAU). All QAU findings at Test Sites must be communicated to the Lead QAU, Study Director, and management at the Testing Facility. The Lead QAU at the Testing Facility may also inspect study phases at Test Sites.

SUBPART C – FACILITIES

This subpart lists the minimum requirements a facility must have for running a GLP study. Figure 2.2 is a floor plan for a small CRO that has the minimum facilities required by the GLPs.

Section 58.41 (General)

Each testing facility must be designed in a manner that allows the proper conduct of the study. There are no hard and fast rules as to what is a correct design. There must be appropriate environmental controls and monitoring procedures for all relevant areas (e.g. animal rooms, test article storage, laboratory areas, archives, computerized operations, etc.).

Section 58.43 (Animal Care Facilities)

The following requirements are needed for maintaining animals on study:

1. A sufficient number of animal rooms or areas to separate species, isolate individual projects, quarantine animals, and house the animals.
2. Separate areas must be provided for diagnosing, treating, and controlling diseased animals.
3. Facilities must exist for collection and disposal of animal waste.

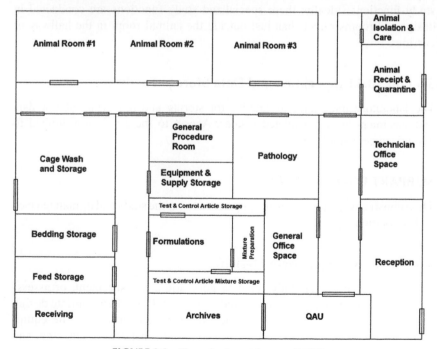

FIGURE 2.2   Floor plan layout for a small CRO

Section 58.45 (Animal Supply Facilities)

Storage areas for feed, bedding, supplies, and equipment are needed. Most importantly, storage areas for feed and bedding must be separated from areas housing animals. A common procedure for laboratories is to hold some feed in the animal rooms in properly labeled and sealed containers. This is rarely a finding on inspections even though technically this should not occur.

Section 58.47 (Facilities for Handling Test and Control Articles)

In order to prevent contamination or mix ups, there must be separate areas for:

1. Receipt and storage of the test and control articles.
2. Mixing of the test and control articles with vehicle.
3. Storage of the test and control article mixtures.

Storage areas for the test and control articles must not be in animal rooms and they must preserve test and control article and mixture identity, strength, purity, and stability.

Section 58.49 (Laboratory Operation Areas)

Separate areas must be provided, as needed, to conduct other necessary study functions. Ideally, these specialized study functions are conducted in designated rooms rather than just outside the animal room in the hallway or corridor.

Section 58.51 (Specimen and Data Storage Facilities)

The laboratory must have an archive for storing and retrieving all raw data and specimens from completed studies. Access to the archives is limited to authorized personnel only.

SUBPART D – EQUIPMENT

This subpart covers the design of equipment used in the study and its maintenance and calibration.

Section 58.61 (Equipment Design)

Equipment used in the study for data generation, measurement, or assessment and equipment for environmental control shall be of appropriate design and adequate capacity to function according to the protocol. The equipment must also be suitably located for operation, inspection, cleaning, and maintenance.

Section 58.63 (Maintenance and Calibration of Equipment)

Equipment must be inspected, cleaned, and maintained and equipment used for data generation must be adequately tested, calibrated and/or standardized. For example, a balance used to collect body weights must be verified for accuracy each time it is used using a standardized weight set. Calibration of equipment to ensure proper function is done less frequently, typically biannually or annually. For example, the standardized weight set should be certified by an outside source annually.

Standard operating procedures must provide details on equipment inspection, cleaning, maintenance, testing, calibration and/or standardization and what to do in the event of equipment failure. The standard operating procedures must also designate who is responsible for the performance of each operation.

Written records must be maintained of equipment inspection, maintenance, testing, calibration, and/or standardizing operations, both routine and non-routine. It is a good idea for each piece of equipment to have a binder that includes: Owner's Manual, Accuracy Verification Log, and Routine/Non-Routine Maintenance Log. This documentation is considered to be "Facility" records rather than "Study Specific" records.

Computer Systems (21 CFR Part 11)

The following procedures and documentation must exist:

1. Validation study (i.e. test scripts), validation plan, and documentation of the plan's execution and results.
2. Maintenance of equipment, including storage capacity and back-up procedures.
3. Control measures over changes made to the computer system (i.e. change control for hardware and software).
4. Evaluation of test data to assure that data are accurately transmitted from analytical equipment to the computer system.
5. Emergency power and back-up procedures.

SUBPART E – TESTING FACILITIES OPERATION

This subpart covers the basic infrastructure elements of running a facility that is conducting a GLP study.

Section 58.81 (Standard Operating Procedures [SOPs])

Written SOPs must be established that cover study methods. An important point is that procedures in the approved protocol or amendments supersede SOPs. All deviations from SOPs and the protocol must be authorized by the Study Director and be documented in the raw data. As mentioned previously, the term "authorization" is confusing since many deviations are made by mistake without appropriate authorization. In these cases, the deviations must be acknowledged by the Study Director, their impact on the study assessed, and a plan of action to prevent similar deviations implemented.

SOPs must be established for at least:

1. Animal room preparation.
2. Animal care.
3. Receipt, identification, storage, handling, mixing, and method of sampling of the test and control articles.
4. Test system observations.
5. Laboratory tests.
6. Handling of animals found moribund or dead during the study.
7. Necropsy of animals or post-mortem examination of animals.
8. Collection and identification of specimens.
9. Histopathology.
10. Data handling, storage, and retrieval.
11. Maintenance and calibration of equipment.
12. Transfer, proper placement, and identification of animals.

Each laboratory area shall have readily available the SOPs relevant to the study functions conducted in that area and only the most current versions of the SOPs must be used. A historical file of SOPs and all revisions must be maintained. Procedures must be in place for familiarizing personnel with SOPs. The use of outdated SOPs is a common finding on inspections despite the extreme measures some laboratories employ to ensure proper change control for their established SOPs.

Differences from the OECD GLPs

Most laboratories will have SOPs that cover the minimum requirements of both the FDA and OECD GLPs since they are all integral for running an animal study. However, the OECD lists the following SOPs that are not mentioned by the FDA GLPs:

- Apparatus use, maintenance, cleaning, and calibration.
- Computerized systems validation, operation, maintenance, security, change control, and back-up.
- Test system (animal) receipt, transfer, proper placement, characterization, identification, and care.

Section 58.83 (Reagents and Solutions)

All reagents and solutions in the laboratory areas must be labeled with their identity, titer or concentration, storage requirement, and expiration date. Expired and deteriorated supplies must not be used on a study. The improper labeling and use of expired reagents and solutions are common findings on inspections.

Section 58.90 (Animal Care)

The animal care SOPs must cover the housing, feeding, handling, and care of animals. All animals received from outside sources (e.g. an animal vendor) must be isolated and their health status evaluated according to acceptable veterinary medical practice.

At the start of the study, animals must be free of any disease or condition that might interfere with the study and, if an animal contracts a disease or condition, the animal must be isolated and treated if the treatment does not interfere with the study. All diagnoses, authorization to treat, and description and date of each treatment must be authorized by the Study Director and documented. For most studies, a Veterinarian will assess the health of the animals during the acclimation period (e.g. detailed physical examinations) and then release only healthy animals for use on the study.

All warm-blooded animals, except suckling rodents, must be appropriately identified. In addition, the identification information must appear on the outside of the housing unit. Typical methods of identifying animals include tattoos, ear tags, and RFID chips and the identification information (i.e. animal number) must also appear on the outside of the animal cage (e.g. cage card).

Animals of different species should be housed in separate rooms. Animals of the same species but used in different studies should also be housed in separate rooms; however, mixed housing can be used if there is adequate differentiation between the studies by space and identification. As a Contracting Scientist, unless there is a reason to group multiple studies in a single room, you should always insist that your study has its own animal room and not be mixed with other studies due to the increased potential for cross-contamination between studies and confidentiality issues.

Animal cages, racks, and equipment must be cleaned and sanitized at appropriate intervals. The actual interval is left up to the laboratory. The interval may also change for different studies depending on the needs of the study. For example, a test article might cause profound diarrhea, necessitating more frequent cage cleaning.

Feed and water used for the animals shall be analyzed periodically to ensure that no contaminants are present that could interfere with the study and the results must be maintained as raw data. The actual interval for analysis and what contaminants are assessed are left up to the laboratory. Common assessments include bacterial contamination, heavy metals, and pesticides. Most feed manufacturers analyze each batch of feed produced and the results of these analyses are usually sufficient for these assessment purposes.

Bedding must not interfere with the study and shall be changed as often as necessary to keep the animals dry and clean. Some types of bedding can cause unwanted effects in the animals. For example, pine chip bedding can induce the level of various drug metabolizing enzymes that could alter the toxicity of the test article.

Pest control materials must be documented and only cleaning and pest control materials that do not interfere with the study must be used.

Differences from the OECD GLPs

Although almost all studies use an acclimation period prior to study initiation, the OECD GLPs specifically state that an acclimation period should be used. In addition, records of the source, date of arrival, and condition of the animals must be maintained, even though this is typically done for FDA GLP compliant studies.

SUBPART F – TEST AND CONTROL ARTICLES

This subpart covers test article receipt, distribution, handling, characterization, storage, mixing, and testing.

Section 58.105 (Test and Control Article Characterization)

1. The identity, strength, purity and composition of the test or control article must be determined and documented for each batch. In addition, methods of manufacturing the articles must be documented. This can be a contentious issue with some CROs since they will insist on having this information on file before initiation of a GLP study. The protocol should clearly identify who and when they will provide and/or maintain this information.
2. The stability of the test and control articles must be determined either before study initiation or during the study.
3. Each storage container must be appropriately labeled and stored under conditions that maintain the article integrity. Storage containers must be assigned to a given test article for the duration of a study.
4. Reserve samples of the raw/bulk test article must be retained for studies that are more than four weeks long. In addition to the raw/bulk test article reserves, the protocol may specify the retention of formulated samples, especially if there are existing formulation issues.

Section 58.107 (Test and Control Article Handling)

Procedures must exist for:

1. Proper storage of articles.
2. Distribution of the articles in a manner that prevents contamination, deterioration, or damage.
3. Identification is maintained during distribution.
4. Receipt and distribution of each batch is documented including the date and quantity of each batch distributed or returned (i.e. chain of custody from receipt to return or destruction).

Section 58.113 (Mixtures of Articles with Carriers)

For each test or control article that is mixed with a carrier (vehicle), the following need to be determined periodically. The actual timeframe is driven by the laboratory and the Sponsor as well as the duration and needs of the study:

1. Uniformity of the mixture. A typical method of doing this is to analyze the mixture at different locations within the container (e.g. top, middle, and bottom) to ensure the concentration of the article is the same at each level.
2. Concentration of the article in the mixture.
3. Stability of the mixture.
   a. Stability can be determined before study initiation or during the study by periodically analyzing the mixture. If analyzed during the study, the period of analysis must cover at least the duration between the preparation of mixtures. For example, if the mixture is prepared weekly, the stability must be assessed over at least a week and under the same conditions of storage/use for administration.

If any mixture component has an expiration date, it must be listed on the container. If multiple expiration dates exist, the earliest one must be used.

SUBPART G – PROTOCOL FOR AND CONDUCT OF A NONCLINICAL LABORATORY STUDY

This subpart covers the minimum content of a protocol and information on the general conduct of a study.

Section 58.120 (Protocol)

Each study must have an approved written protocol that indicates the objectives and all methods for the conduct of the study. It is important to note that the protocol is different from a general research plan that might be submitted to support a grant application. The GLP study protocol provides clear direction on how to conduct the given study and all key study functions. In contrast, a research plan typically covers many different studies in very generic terms, with the individual study conduct details left up to the investigator.

Specific components of the protocol are listed in this section. In general, most whole animal studies will contain all of these sections and possibly additional sections (see the FDA GLPs for a complete listing of all the elements). A few key points:

1. The protocol must include the date of approval by the Sponsor. Although the GLPs do not require written approval of the Sponsor, many laboratories will not consider the protocol approved until signed by the Sponsor. The protocol is a contract between the Sponsor and the laboratory and ensures that the laboratory conducts the study according to the Sponsor's requirements. If the

Sponsor does not sign the protocol, it is important that the date of Sponsor approval is supported by some sort of documentation (e.g. phone log, e-mail, memo, letter from Sponsor).

2. The Study Director must sign and date the protocol.
3. All changes or revisions to an approved protocol and the reasons for the changes must be documented, signed by the Study Director, dated, and maintained with the protocol. These are often referred to as protocol amendments. Although not specifically required by the GLPs, the Sponsor should approve and sign all protocol amendments before they are implemented. If verbal or written (e.g. e-mail) approval is given in lieu of a signature due to logistical issues (e.g. the Sponsor is traveling), the transcript or e-mail should be included with the amendment or study data and the amendment signed later by the Sponsor.

Section 58.130 (Conduct of a Nonclinical Laboratory Study)

This section covers some basic, but essential, components of running a GLP study. Deviation from these is a common finding by inspectors and can result in GLP deviations and warning letters.

1. The study must be conducted in accordance with the protocol. This essentially means that you must do what the protocol (and amendments) states, no more and no less.
2. The test systems must be monitored according to the protocol. For example, if the protocol states to conduct clinical observations on the animals twice a day, this must be done twice a day and not once a day or three times a day.
3. Specimens must be identified with the following:
   a. Test system
   b. Study
   c. Nature (i.e. what is the specimen, such as liver or heart tissue, 4h pharmacokinetic sample)
   d. Date of collection
   e. Animal number or other unique identification. Surprisingly, this is not a specific requirement of the GLPs but is essential for accurately identifying which animal the samples came from.
4. Records of gross findings for a specimen should be available to the pathologist so that they can correlate the findings with histopathological observations. This is not an absolute requirement but it is a highly recommended practice, even for non-GLP studies since it allows a clear association between gross lesions and histopathological correlates.
5. Data generation – this is an essential aspect of the GLPs.
   a. All data generated must be: Attributable, Legible, Contemporaneous, Original, and Accurate (ALCOA). The GLPs state that data generated must be recorded:

    i. Directly,

    ii. Promptly (i.e. data cannot be recorded at a later time), and

    iii. Legibly in ink (this prevents erasing the data), except for electronic-captured data.

b. All data entries must be dated on the date of entry and signed or initialed by the person entering the data. Backdating is never acceptable. It is always better to date with the current date and provide an additional explanation if needed. An accurate date is essential to study reconstruction, which is a critical aspect of the GLPs.

c. Any change in entries:

    i. Must not obscure the original entry

    ii. Must indicate the reason for the change, and

    iii. Must be dated and signed at the time of the change.

d. For automated data collection systems:

    i. The individual entering the data must be identified,

    ii. Any change in entries

        1. Must be made so as not to obscure the original entry

        2. Indicate the reason for the change

        3. Must be dated, and

        4. The responsible individual for making the change identified.

    iii. The laboratory must be careful with how changes to original entries are initiated and processed, especially when authority to make changes is limited to supervisory or IT personnel. Timeliness of revisions may come into question if changes are made days, weeks, or years after the original observations. There should be supporting documentation requesting such changes. This documentation should include a justification for making the change.

SUBPART J – RECORDS AND REPORTS

This subpart covers the content of a study report and storage and retrieval of records.

Section 58.185 (Reporting of Nonclinical Laboratory Study Results)

A final report must be prepared for each study. This section provides the required elements of the report (see the FDA GLPs for a complete listing of all the elements). Several important points are that the report must:

1. Describe all circumstances that may have affected the quality or integrity of the data. These are essentially the deviations to the study protocol, amendments, and SOPs that the Study Director determines were serious enough possibly to affect the integrity of the study. Many deviations are minor and do not affect the study; however, many laboratories choose to list all deviations, regardless of their impact on the study. For the ones that are serious enough possibly to affect the integrity of the study, the Study Director needs to provide details about the potential impact on the study.

2. Include the signed and dated reports of individual scientists or other professionals involved in the study.

3. List the location where all specimens, raw data, and the final report are to be stored.

4. Contain the QA statement. A key point is that the QA Statement as required by 21 CFR 58.35(b)(7) and 58.185(a)(14) is not a statement of GLP compliance. Only the Study Director can make such a statement [21 CFR 58.33(e)]. The QA Statement is often confused with the GLP Compliance Statement that is required by the product application submission process.

5. Be signed and dated by the Study Director.

Corrections or additions to the final report must be in the form of an amendment by the Study Director and must clearly identify the part that is being added to or corrected and the reasons for the correction or addition. The corrections or additions must be signed and dated by the person responsible. The finalization process for a report amendment is usually the same as for the final report. The QAU will audit for accuracy and compliance to the GLPs, protocol, and applicable SOPs. The QAU will normally issue an additional QA Statement to reflect the review of the amended report.

Section 58.190 (Storage and Retrieval of Records and Data)

1. All raw data, documentation, protocols, final reports, and specimens (that are amendable to storage) must be indexed and retained in an archive that has appropriate environmental controls.

2. An individual must be identified as being responsible for the archives and only authorized personnel may enter the archives.

Section 58.195 (Retention of Records)

This section outlines the retention period for various types of study records and specimens.

OECD Multisite Studies

The OECD GLPs provide specific guidance on running multisite studies under GLPs and some of this information was covered in the previous sections. Essentially, a multisite study consists of the Study Director, Testing Facility Management, Testing Facility QAU, Test Sites, Principal Investigator(s), Test Site Management, Test Site QAU, and all the various personnel at each entity performing study functions. The OECD GLPs provide strong warnings for conducting multisite studies since they greatly increase the complexity of running GLP studies and ensuring that all phases are conducted in compliance with the GLPs. These warnings should not be taken lightly since not only are the logistical issues of coordinating the various study functions increased but you often have to contend with varying opinions on how to handle different GLP-related issues at the different sites. The most important thing to keep in mind with multisite studies is that it is ultimately the Study Director who is responsible for GLP compliance and overall conduct of the study. Figure 2.3 is a diagram of the reporting structure for a typical multisite study where the Testing Facility (main test site) is where the Study Director is located and is running the in-life portion of the animal study (i.e. dosing, observations, necropsy, and sample collection). One Test Site is enrolled that is analyzing blood for hormone levels and a second Test Site is enrolled that is conducting immunohistochemistry analysis of liver sections collected during necropsy. The solid lines represent direct reporting lines. The dotted lines represent links between personnel that involve indirect oversight and/or exchange of critical information (e.g. Test Site QAU informing the Testing Facility Study Director of relevant findings).

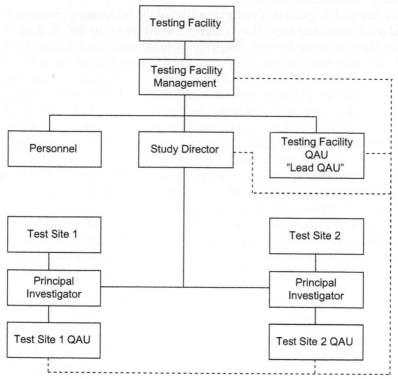

FIGURE 2.3   Reporting structure for a typical multisite GLP study

GLP FACILITY INSPECTIONS

Test Facilities conducting FDA GLP studies are audited by the FDA to ensure that studies are being conducted in compliance with the GLPs and that the study report accurately reflects what was done on the study and matches the raw data. There are four types of inspections:

1. GLP laboratory inspection (also called surveillance inspection).
2. Data audit inspection.
3. Directed inspection.
4. Follow-up inspection.

GLP laboratory inspections are periodic, routine determinations of a laboratory's compliance with GLP regulations. These inspections include a facility inspection and audits of on-going and/or recently completed studies. Studies may be audited but the actual studies may be determined by the inspector instead of being specifically determined prior to the inspection, which occurs during a data audit inspection. While these inspections are not "GLP Qualifying" inspections, they do play an important role in ensuring that the laboratory understands that the FDA GLPs are not merely guidelines, but that they are actually enforceable regulations.

A data audit inspection is made to verify that the information contained in a final report submitted to the FDA is accurate and reflected by the raw data. This type of inspection involves a thorough review of the selected study(s) and all the raw data associated with the study. The actual studies that will be audited are determined by the FDA Center which is reviewing the product application (e.g. the Center for Drug Evaluation and Research [CDER] for a New Drug Application). This reinforces why it is important that laboratories have good archival procedures in place. A detailed index can be instrumental in being able to locate requested data in a timely manner. Being able to produce requested data quickly can set the tone of the inspection.

Directed inspections focus on a specific need of the agency and are conducted for various compelling reasons including but not limited to questionable data in a final report, tips from informers, etc.

A follow-up inspection is conducted after a previous GLP inspection revealed questionable practices and conditions (e.g. a Warning Letter was issued to the laboratory for GLP deviations). The follow-up inspection checks to see if proper corrective actions have been taken by the laboratory to remediate their deficiencies.

GLP laboratory inspections are typically conducted at least every two years; whereas, the other types of inspections are conducted as needed.

During an inspection, the inspector will follow the guidance in the FDA Compliance Program Guidance Manual; Chapter 48 – Bioresearch Monitoring (http://www.fda.gov/downloads/ICECI/EnforcementActions/Bioresearch-Monitoring/ucm133765.pdf). Inspections may or may not be pre-announced.

Generally, GLP laboratory inspections are pre-announced; whereas, the other three types of inspections are often not pre-announced. Laboratories typically have an SOP that covers regulatory inspections and this should be reviewed during your audit of a laboratory to ensure that they have adequate procedures in place. It is also important to question management, Study Director(s), and the QAU on how they handle inspections since you can get a good feel for their general approach. You should also ensure that their procedures include notifying you, the Contracting Scientist, of any inspections as soon as possible so that you can take any needed action. Often, companies have procedures to deal with GLP inspections, particularly data audit inspections of the company's studies. However, by the time an inspection occurs, there is often little the Contracting Scientist can do but keep their fingers crossed, so it is best to ensure that the laboratory and the studies are fully GLP compliant prior to any inspection since the Sponsor has the most to lose if a GLP study goes awry (e.g. time, money, delay to market).

At the conclusion of an inspection, the inspector will issue an Establishment Inspection Report (EIR) and classify it as:

- No Action Indicated (NAI) – no objectionable conditions or practices found. This is the best classification and attests to a laboratory's excellent compliance with the GLPs.
- Voluntary Action Indicated (VAI) – objectionable conditions or practices found but no administrative or regulatory action will be taken or recommended. GLP issues were found but they were not serious enough to warrant a warning letter.
- Official Action Indicated (OAI) – regulatory and/or administrative actions will be recommended. The outcome of this will typically be a Warning Letter (FDA 483), which identifies serious GLP deviations.

When scoping out a new laboratory, you should ask for copies of past EIRs and Warning Letters that may have been issued to the laboratory. If they are unwilling to provide you with copies, they may be trying to cover up issues identified during past inspections. You can always request copies of EIRs and Warning Letters via the US Freedom of Information Act; however, this can take a while. You can search specifically for Warning Letters via the FDA website at: http://www.fda.gov/ICECI/EnforcementActions/WarningLetters/default.htm.

Although Warning Letters are a major red flag, sometimes there are acceptable reasons why a Warning Letter was issued. For example, a CRO was issued a Warning Letter for having inadequate QAU procedures. After reviewing the CRO's QAU procedures, it was determined by the Contracting Scientist that the CRO's procedures were more encompassing than another CRO he was working with that never had an FDA 483 issued against them. In addition, the CRO altered their QAU procedures to address the concerns in the FDA 483 and a subsequent audit of the laboratory found the new procedures acceptable. Therefore, the Contracting Scientist decided to use this laboratory.

European Union (EU) Facility Inspections

The EU takes a different approach than the USA when it comes to inspecting facilities for GLP compliance. In the USA, any laboratory can conduct a GLP study and claim that the study complies with GLP regulations. It is not until the laboratory or a specific study is audited by the FDA that the final determination of GLP compliance is made. In contrast, laboratories in the EU must undergo inspection and receive certification before they can conduct a GLP study. Once the GLP approval is given to the laboratory, the general assumption is that any GLP study coming out of that laboratory is fully GLP compliant or at least the laboratory has the basic infrastructure that is capable of being GLP compliant. The laboratories undergo periodic renewal inspections but, in contrast to the USA, study directed inspections are not conducted as frequently. Both approaches have pros and cons. Regardless of the approach, it is important for the Contracting Scientist to realize the differences in the GLP compliance procedures, especially when a study conducted in one country is being submitted to support regulatory approval in a country that has a different GLP compliance approach. For example, we conducted a study at a GLP certified CRO in an EU country to support drug approvals in the USA and the EU. The study coming out of the EU country was considered fully GLP compliant by the EU regulatory authorities since it came from a GLP certified laboratory and the study was accepted in support of the drug approval. In contrast, upon auditing the study, the US FDA found major GLP deviations with the study and determined that it could not be used to support the drug registration in the USA.

GLP AUDITING CHECKLIST

The following checklist can be used to scope out a new laboratory's GLP compliance procedures or audit an existing laboratory to ensure they continue to have appropriate GLP procedures in place and are conducting your study in compliance with GLPs. In general, the checklist follows the order of the GLPs. A thorough audit can take several days and should involve the review of critical SOPs; interviews with key study personnel, QAU, and management; training records review; review and critique of the facilities and operations; assessment of test and control article and mixture receipt, handling, storage, and distribution procedures; observation of critical study functions; auditing of protocol development and reporting procedures; and assessment of data recording, including automated data capture systems, and archiving. When going through the checklist, if any of the answers are "No", those specific areas require more in depth review of the laboratory's procedures to ensure they are in compliance with GLP practices. While some of the items listed on the checklist are not specifically required, they may be key elements to demonstrate GLP compliance, especially when/if the FDA audits the laboratory and/or study.

FACILITY ORGANIZATION & PERSONNEL	YES	NO	N/A
PURPOSE OF THIS SECTION: • Determine if the organizational structure is appropriate to ensure that studies are conducted in compliance with GLP regulations • Determine whether personnel, management, study directors, and the quality assurance unit (QAU) are meeting their responsibilities under GLPs			
Personnel (21 CFR 58.29) Personnel must be qualified for their assigned duties and appropriate documentation of their qualifications and training must be maintained			
1. Curricula vitae (CV) are available and current for all employees and they are signed and dated			
2. CVs provide adequate detail of education, past experience, and formalized training/meetings attended			
3. CVs are maintained after departure of personnel			
4. Job descriptions are available and current			
5. Job descriptions are "job" specific, not "person" specific			
6. Job descriptions include minimum education and experience requirements			
7. Job descriptions are signed and dated by management and historical copies are maintained			
8. Training records are available and current. Should note in the inspection report which records were reviewed for future reference			
9. Training records are reviewed periodically. How often and by whom? Is there an SOP for training records?			
10. Personnel are adequately trained and qualified for their assigned duties			
11. Personnel have been trained in GLPs How often is refresher training provided?			
12. Signatures and initials of study personnel are on file How often are they updated?			
13. Sanitation precautions taken and appropriate clothing worn to prevent contamination of test and control articles and animals			
14. Personal illnesses are reported and action taken to avoid contamination of test and control articles and animals			
15. Sufficient number of qualified personnel for timely and proper conduct of study How many total personnel at facility and how many studies are run?			
Management (21 CFR 58.31) Management is responsible for providing resources, personnel, and infrastructure for running a GLP study. They must play an active role in running the facility			
1. Review the organization chart and determine the various organization units, their role in carrying out GLP study activities, and the management responsible for the units including off-site personnel Is the organization chart current and dated?			
2. Historical organization charts are maintained			

FACILITY ORGANIZATION & PERSONNEL	YES	NO	N/A
3. Designates a study director before the study begins An SOP covers the process of assigning a Study Director			
4. Controls Study Director workload – review the master schedule to determine the number and types of studies per Study Director Does the master schedule list non-GLP studies so that you can adequately assess Study Director workload?			
5. Assures that sufficient personnel, resources, facilities, equipment, materials and methods are available as scheduled			
6. Assures that a Quality Assurance Unit (QAU) is present, functioning appropriately, and independent from the studies they are reviewing (i.e. they do not direct or conduct study functions for the study they are auditing) Who does the head of the QAU report to?			
7. Assures that any deviations or deficiencies reported by the QAU are communicated to the Study Director and appropriate study personnel and corrective actions are taken What is the reporting and communication process?			
8. Assures that test and control articles or mixtures are appropriately tested for identity, strength, purity, stability, and uniformity. If test article characterization and stability is performed by the Sponsor, verify that the test facility has received documentation that this testing has been conducted			
9. Assures that study personnel know and follow any special test and control article handling and storage procedures			
10. Reviews and approves protocols. This is not specifically required under the GLPs but is a good practice How is approval documented?			
11. Reviews and approves standard operating procedures (SOPs) How is approval documented?			
12. Provides GLP and appropriate technical training			
13. Assures that personnel clearly understand the functions they are to perform			
Study Director (21 CFR 58.33) The Study Director is the single point of control of the GLP study and must play an active role in its conduct to ensure its successful completion			
1. The Study Director is actively involved with the study. Review study-related communications, room visit logs, data review records, etc.			
2. The Study Director is immediately notified of any problems that may affect the quality and integrity of the study			
3. Documents unforeseen circumstances that may affect the quality and integrity of the study and implements corrective action. Check the laboratory's procedures for reporting and acting upon protocol deviations How and when are deviations reported to the Study Director? Is the impact on the study and corrective actions documented by the Study Director?			

FACILITY ORGANIZATION & PERSONNEL	YES	NO	N/A
4. Assures that the protocol and any amendments are approved and followed			
5. The Study Director actively communicates with the lab and vice versa How are the communications documented (e.g. e-mail, phone log, Study Director/Laboratory Communication logs)?			
6. Assures that all data are accurately recorded and verified – check to see if Study Director has reviewed raw data and how frequently			
7. Assures that data are collected according to the protocol and SOPs			
8. Assures that study personnel are familiar with and adhere to the study protocol and SOPs			
9. Assures that study data are transferred to the archives at the close of the study How soon after completion does the transfer take place? Is there an SOP that covers the transfer?			
10. All Study Director communications (e.g. e-mail, phone logs, faxes) are archived with the study			
11. When the Study Director is on leave, the laboratory assigns an "acting" or "deputy" Study Director and this is documented in the raw data			
Quality assurance unit (21 CFR 58.35) The QAU must be independent from personnel conducting a GLP study and assures management of GLP compliance. The QAU monitors significant study events and facility operations and reviews records and reports for GLP compliance			
1. QAU is in place Is it on site or contracted?			
2. QAU is separate from and independent of personnel engaged in the direction and conduct of the study			
3. QAU reports directly to management			
4. QAU appears to have management support			
5. SOPs exist for the following QAU activities: QAU duties and responsibilities QAU training procedures Maintenance of a master schedule sheet Maintenance of copies of all protocols and amendments Scheduling of in-process inspections and audits Inspection of study at intervals adequate to ensure the integrity of the study and maintenance of the inspection records Immediate notification of the Study Director and management of any problems that may affect the integrity of the study Submission of periodic status reports on each study to the Study Director and management Auditing procedures (protocol, critical phase, raw data, final report) Reporting procedures for inspections and audits Preparation of a statement in the final report that specifies the dates inspections were made and findings reported to the Study Director and management Inspection of computer operations			

FACILITY ORGANIZATION & PERSONNEL	YES	NO	N/A
6. Maintains copies of all signed, approved protocols, amendments and deviations			
7. Critical phase inspections are conducted for each study How frequently are they typically conducted and what phases are typically inspected? Inspections are left to the discretion of the QAU Who is responsible for ensuring that audits are completed?			
8. Maintains records of inspection dates, study inspected, phase inspected, name of inspector, and written inspection procedure			
9. How much time does the QAU spend on: In-process inspections Final report audits Is this sufficient to detect problems in critical study phases and if there are adequate personnel to perform study functions? Does the QAU have adequate time allowed to conduct their audit or are they often pressured by deadlines?			
10. Conducts periodic facility inspections. How often are they conducted? Are findings reported to the Study Director and management?			
11. QAU has GLP training			
12. Provides periodic GLP training to facility personnel How often is refresher training taken?			
13. QAU records are kept in a central location separate from other study files			
14. QAU is adequately staffed How many personnel are in the QAU?			
15. Maintains copies of relevant GLPs, advisories, and guidelines (as applicable)			
16. Reviews the final study report and raw data package to assure that the report: Accurately summarizes the findings and matches the raw data Describes the methods and SOPs that were followed and any deviations that occurred What percentage of raw data does the QAU inspect? What percentage of the report data is audited against the raw data?			
17. If the study involves an interim report, the interim report is reviewed by the QAU in the same manner as a final study report			
18. Prepares and signs a statement to be included with the final study report which specifies the dates inspections were made and findings reported to management and the Study Director			

FACILITY ORGANIZATION & PERSONNEL	YES	NO	N/A
19. Maintains copy of the master schedule. The master schedule may be generated by another group (e.g. Test Facility Management); however, the QAU must maintain a copy Who generates the master schedule? Contains sponsor identity (or code), name of Study Director, test substance, test system, nature of study, initiation date, current status Status of each study is accurate Master schedule can be indexed by test article Historical master schedules are maintained Master schedule lists non-GLP studies			
20. Studies are entered on master schedule after protocol is signed by the Study Director (i.e. study is initiated)			
21. Studies are removed from master schedule after final report is signed by the Study Director			

FACILITY	YES	NO	N/A
PURPOSE OF THIS SECTION: • Determine if the facilities are of adequate size and design and are adequately maintained and monitored			
General (21 CFR 58.41)			
1. Testing facility is suitable for the conduct of the study			
2. A floor plan is available and current Take a copy of the floor plan on the facility inspection to ensure it is accurate			
3. Work areas are of adequate size, location, and construction			
4. Work areas are secure from the outside What are the security procedures?			
5. Work areas are neat, clean and uncluttered			
6. Relevant SOPs are available in work areas			
7. Disposal facilities are adequate			
8. Facility is clean Review SOP on cleaning critical areas and equipment to ensure cleaning materials do not interfere with study			
9. When two or more functions are conducted in a single room, is adequate separation maintained when needed?			
10. Are environmental controls and monitoring procedures adequate for: Animal rooms Test article storage areas Laboratory areas Handling of biohazardous materials Feed storage area			

EQUIPMENT	YES	NO	N/A
11. Equipment used to generate, measure or assess data is adequately tested, calibrated, and/or standardized before use on study			
12. Balances/scales: Balances/scales and/or weights are certified at adequate intervals Accuracy verification is performed before each weighing			
13. HVAC system design and maintenance is adequate for critical areas and includes documentation Filter changes are documented Temperature/humidity is monitored HVAC intakes are away from cars, fume hood emission ports, etc.			
14. Non-dedicated equipment for the preparation of test and control article mixtures is cleaned and decontaminated to prevent cross-contamination			
TESTING FACILITY OPERATIONS	YES	NO	N/A
PURPOSE OF THIS SECTION: • Determine if the facility has appropriate SOPs and follows them • Determine if the facility ensures the quality of reagents and solutions • Determine if animal care and housing minimizes stress and uncontrol led influences on animals			
Standard operating procedures (21 CFR 58.81)			
1. The following GLP SOPs must exist. For each SOP, a single SOP may be used or they may be divided among multiple SOPs as long as all the areas are covered: Animal room preparation Animal care Receipt, identification, storage, handling, mixing, and method of sampling and analysis of the test and control articles and mixtures, including chain of custody procedures Test system observations Laboratory tests Handling of animals found moribund or dead during study Necropsy of animals or postmortem examination of animals Collection, preparation, identification, and transfer of samples and specimens, including storage and chain of custody procedures Histopathology Data handling, storage, and retrieval Maintenance and calibration of equipment, including remedial actions and the person responsible Transfer, proper placement, and identification of animals			

TESTING FACILITY OPERATIONS	YES	NO	N/A
2. The following SOPs should also be present:			
Study Director assignment including roles and responsibilities			
Replacement of a Study Director			
Validation of computer systems used to enter, store, manipulate and report data			
Writing, reviewing, revising, and retiring SOPs			
Archive procedures and the personnel responsible			
Duration of archive period			
Archivist duties and responsibilities			
Archives indexing and maintenance procedures			
Training records and CVs – review and update frequency and approval procedure			
Data recording, quality and integrity, including input (data checking and verification), output (data control), and audit trail covering all data changes (error corrections, methods, codes, and rounding)			
Preparation of standard solutions			
Assigning expiration dates to reagents and solutions			
Labeling of reagents and solutions			
Cleaning of glassware			
Preparation, review, and approval, and amending the final report			
3. An index of the currently approved SOPs is available			
4. Complete set of the SOPs, including revisions, is maintained in the archives			
5. Historical SOPs are retained in the archives and can be readily accessed			
6. SOPs have proper authorization signatures and dates			
7. SOPs are clear, complete, and can be followed by a trained individual			
8. SOPs are periodically reviewed to ensure they are still applicable and are representative of the actual procedures in use How often is this review conducted and by whom?			
9. Study personnel follow the relevant SOPs. It is helpful to observe study personnel conduct SOP-driven functions and assess their compliance with the procedures			
10. For deviations from SOPs, the Study Director assesses the impact on the study and documents the deviation in the raw data			
11. Changes to SOPs are authorized in writing by management			
12. Personnel are familiar with the location and content of SOPs that are applicable to their duties			
13. Only currently approved SOPs are in use (no superseded or rescinded SOPs in use). It is helpful to check the SOPs at work areas to ensure only the currently approved SOPs are present			
14. Who is responsible for maintaining the SOP system? How do they ensure that only currently approved SOPs are in use?			
15. The review of SOPs that are relevant to a person's duties is documented in their training file			

TESTING FACILITY OPERATIONS	YES	NO	N/A
Reagents and solutions (21 CFR 58.83)			
1. What are the procedures to purchase, receive, label, and determine the acceptability of reagents and solutions?			
2. All reagents and solutions in the laboratory areas are labeled with identity, titer or concentrations, storage requirements, and expiration date			
3. No expired reagents or solutions are in use			
4. Profile data accompanying each batch of control reagents are used for automated analytical equipment			
Animal care (21 CFR 58.90)			
1. Adequate SOPs are in place covering the environment, housing, feeding, handling, and care of laboratory animals Are the SOP instructions being followed?			
2. Newly received animals are appropriately isolated and their health status evaluated and documented			
3. Before assignment to study, animals are free of disease or any condition that might interfere with the study. Who conducts this assessment, when is it conducted (e.g. during acclimation), and what is the procedure for releasing animals for study?			
4. Animals are appropriately identified (e.g. tattoo, RFID chip, ear tag, etc.)			
5. Animal identification is also present on the outside of the cage Verify that the animal ID matches the cage ID for several animals			
6. Different species are housed in separate rooms			
7. If multiple studies using the same species are housed in a single room, there is adequate separation between studies			
8. Treatments given to animals that have become diseased are authorized by the Study Director and documented			
9. The facility has an Institutional Animal Care and Use Committee How often do they meet? Do SOPs cover their operation?			
10. Daily observation logs are maintained Verify that logs contain reports of dead animals and external gross lesions or masses			
11. Cages, racks, and accessory equipment are adequately cleaned and sanitized at appropriate intervals			
12. Bedding used in cages or pens is adequate and changed at a frequency to keep animals dry and clean			
13. Bedding does not interfere with the study (e.g. pine chip bedding may cause unwanted hepatic enzyme induction)			
14. Feed and water are analyzed periodically and analytical documentation is maintained in the raw data Are contaminants present at levels that interfere with the study? Who determines the analyses that are conducted? Who reviews the analyses?			

Chapter | 2   Good Laboratory Practices

TESTING FACILITY OPERATIONS	YES	NO	N/A
15. Are pest control agents or methods used? If yes: Who is responsible for the pest control program? Are the pest control agents and areas of application documented? Pest control agents do *not* interfere with the study Is the pest control applicator escorted during application?			
16. Cleaning materials are appropriately labeled and do *not* interfere with the study			

TEST AND CONTROL ARTICLES	YES	NO	N/A
PURPOSE OF THIS SECTION: • Determine that test and control articles and mixtures meet protocol specifications throughout the course of the study and that accountability is maintained			
Test and control article characterization (21 CFR 58.105)			
1. The identity, strength, purity, and composition (i.e. characterization) of the test and control articles are determined for each batch and are documented			
2. The stability of the test and control articles is documented			
3. If the Sponsor conducts the characterization and stability of the test and control articles, does the test facility have documentation from the Sponsor that this testing has been conducted and does the protocol clearly specify the responsible party?			
4. Test compounds labeled with name, CAS or code number, batch number, expiration date (if any), and storage conditions (where appropriate)			
5. Storage containers are assigned for the duration of a study			
6. Original test substance container kept until the study completion date			
7. Transfer from the point of collection to the analytical laboratory is documented			
8. Reserve samples of test and control articles for each batch are retained for studies >4 weeks			
9. If the feed is used as the control article, a reserve sample of the feed is retained			
Test and control article handling (21 CFR 58.107)			
1. Each storage area has unique identification (e.g. test substance storage or mixture storage to preclude mix-up and contamination)			
2. Storage area is adequately ventilated			
3. Environment of the storage area is monitored and documented Monitoring devices are adequately calibrated			
4. Storage area has limited access			
5. Receipt and condition upon receipt adequately documented			
6. Bulk inventory log maintained for each batch and includes the date and quantity of each batch distributed or returned			

TEST AND CONTROL ARTICLES	YES	NO	N/A
7. Adequate usage/accountability documentation, including chain of custody procedures			
8. Distribution procedures prevent contamination, deterioration, or damage			
9. Transport containers are adequately cleaned			
10. The test and control article handling and identification during the distribution and administration to animals is clear and prevents mix-ups			
Mixtures of articles with carrier (21 CFR 58.113)			
1. Preparation, sampling, testing, storage, and administration of mixtures of test and control articles with carriers are adequate			
2. Mixtures are analytically characterized for: Uniformity (homogeneity) Concentration of the test or control article in the mixture Stability under the study conditions Uniformity and concentration are assessed periodically during the study. Stability can be done prior to or concurrently with the study			
3. Analytical mixture results are reported to the Study Director in a timely manner			
4. If a component(s) of the mixture has an expiration date, the earliest expiration date is listed on the container			
Analytical samples			
1. Adequate specimen, test, control, and mixture log-in procedures			
2. Samples are 100% tracked and documented throughout the study (receipt through analysis) – chain of custody procedures			
3. Storage conditions documented			
4. Samples stored separate from standards			
5. Adequate procedures for disposition of specimens/samples			
PROTOCOL FOR AND CONDUCT OF A NONCLINICAL LABORATORY STUDY	YES	NO	N/A
PURPOSE OF THIS SECTION: • Determine if study protocols are properly written and authorized and that studies are conducted in accordance with the protocol and SOPs			
Protocol (21 CFR 58.120)			
1. Protocol preparation SOPs are followed			
2. Protocol contains the specific elements listed in 21 CFR 58.120(a)(1)–(12)			
3. All changes, revisions, or amendments to the protocol are authorized, the reason for the change specified, signed, and dated by the Study Director			
4. Protocol and amendments are filed together with corresponding protocol records and are readily available to personnel carrying out studies			
5. Distributed copies of approved protocol contain all changes, revisions, or amendments. Check several study protocols to ensure they are complete			

PROTOCOL FOR AND CONDUCT OF A NONCLINICAL LABORATORY STUDY	YES	NO	N/A
Conduct of a nonclinical laboratory study (21 CFR 58.130)			
1. Evaluate laboratory operations, facilities, and equipment to verify compliance with the protocol and SOPs for: Animal monitoring Recording raw data (manual and automated) Corrections to raw data Randomization of animals to study Collection and identification of specimens Authorized access to data and computerized systems			
2. Data generated during the conduct of the study are: *Attributable* – raw data can be traced by signature/initials and date to person observing and recording the data *Legible* (readable, interpretable, and permanent [e.g. in ink]) *Contemporaneous* – raw data must be recorded at the time of the observation *Original* – must be the first recording of the data *Accurate* – raw data are true and complete observations Review the raw data of several studies to ensure the data meet these criteria			
3. For data entry forms, all fields are completed or a written explanation for empty fields is provided			
4. If the person making entries is different than the person making the observations, both names are signed and dated One person may be the observer and one person may be the recorder; however, both individuals and their roles must be clearly documented			
5. Changes in raw data entries: Must not obscure the original entry Indicate the reason for the change Signed and dated at the time of change by the person making the change			
6. Data are well presented and afford adequate study reconstruction All data collected are reported Individual data are reported to support summary tables and figures Individual data tables reflect all audit trails (i.e. show when, how, why, by who data were revised)			
7. Unique study number is recorded on each form used to document data for the study			
8. Specimens are labeled by: Animal number Test system Study Nature (i.e. what is the specimen [e.g., liver tissue, 4h plasma pharmacokinetic sample]) Date of collection Storage conditions If specimen size precludes this amount of information, a code is used and a code sheet is available			

PROTOCOL FOR AND CONDUCT OF A NONCLINICAL LABORATORY STUDY	YES	NO	N/A
9. Records of gross findings for a specimen are available to the pathologist for correlating histopathological findings			

10. Specimen storage areas:
   Clean, organized, and free from contamination
   Limited access
   Control and treated samples are adequately separated
   Specimen/sample inventories maintained for each storage area (including refrigerators and freezers)
   Environment is monitored and documented
   Monitoring devices are calibrated and documented
   Back-up generators (or other systems) available to preserve environment and they are routinely tested
   Alarm system in place in case of malfunction			

RECORDS AND REPORTS	YES	NO	N/A
PURPOSE OF THIS SECTION: • Determine how the test facility stores and retrieves raw data, documentation, protocols, final reports, and specimens			
Reporting of nonclinical laboratory study results (21 CFR 58.185)			
1. A final report is prepared for the study and contains at least the elements listed in 21 CFR 58.185(a)(1)–(14)			
2. Contains signed and dated reports of each of the individual scientists involved in the study			
3. Contains the QA inspection statement			
4. Is signed and dated by the Study Director			
5. Corrections are made in the form of an amendment by the Study Director. The amendment clearly identifies the part being changed, the reason for the change, and is signed and dated by the responsible person			
6. Final report is issued for studies that are stopped prior to completion (e.g. due to power failure, disease outbreak) and it includes the reason for study termination			
7. There is an SOP for the preparation, review, approval, and amending of the final report and it is followed			
Storage and retrieval of records and data (21 CFR 58.190)			

1. Adequate secure archive with limited access
   Who is responsible for the archives?
   Who has access to the archives?
   Is a sign-in/sign-out log maintained?			

2. Archive sufficient to store raw data and specimens			
3. Is an off-site archive used? Is it GLP compliant and can you retrieve the documents and specimens in a timely manner?			
4. Environment of archives is controlled and monitored to prevent deterioration of data and specimens			

5. Archive area is protected from disaster
   Fire suppression system?
   Flood protection?
   Fire/water proof storage?			

RECORDS AND REPORTS	YES	NO	N/A
6. Raw data, documentation, protocols, final reports, and specimens from completed studies are archived How long are perishable specimens retained? Is the Sponsor notified before disposal of any such specimens?			
7. QAU records are stored separately from study records			
8. Equipment/plant maintenance/calibration records are archived (historical records)			
9. Material is indexed to expedite retrieval			
10. Procedures in place for logging data in and out (date in/out, who, why)			
11. Exact copies are archived when originals are shipped (use "certified copy" stamp or cover page)			
12. In-progress study records kept in secure conditions Where are the temporary storage areas?			
COMPUTER OPERATIONS	YES	NO	N/A
PURPOSE OF THIS SECTION: • Determine if computer operations meet GLP and 21 CFR Part 11 requirements			
Personnel			
1. The personnel involved in the design, development, and validation of the computer system is documented			
2. The personnel responsible for the operation of the computer system, including inputs, processing, and data output is documented			
3. Personnel have appropriate training, including GLP training, and it is documented			
4. Personnel using computer systems are appropriately trained			
Facility			
1. Computer operations and archived data are housed under appropriate environmental conditions (e.g. protected from heat, water, and electromagnetic forces)			
Equipment			
1. Validation study exists, including validation plan and documentation of the plan's completion			
2. Procedures exist and are documented for maintenance of equipment including storage capacity and back-up procedures			
3. Control measures over changes made to the computer system, which include the evaluation of the change, necessary test design, test data, and final acceptance of the change			
4. Evaluation of test data to ensure that data are accurately transmitted and handled properly when analytical equipment is directly interfaced with the computer			
5. Procedures for routine and emergency back-up of the computer system			

COMPUTER OPERATIONS	YES	NO	N/A
Testing facility operations			
1. Historical file of outdated or modified computer programs is maintained. If not, a hard copy of all programs is present and stored			
2. High Level System Document is maintained Lists hardware and software (including current versions) used and how equipment (e.g. balances) is interfaced into the overall system			
Conduct of a nonclinical laboratory study			
1. The individual responsible for direct data input is identified at the time of data input			
2. Any change in automated data entries are made so as not to obscure the original entry, indicates the reason for the change, is dated, and the responsible individual is identified How long after the in-life portion of a study is completed can changes/edits be made to the data?			
3. Back-up procedures are in place in case of system failure (e.g. manual data entry forms) Relevant forms are readily accessible and personnel are trained in their use?			
Records and reports			
1. Description of any computer program changes is included in the final validation report			
Archives			
2. How and where are computer data and backup copies stored and indexed?			
3. Environment is controlled and monitored to prevent deterioration			
4. Electronic media has the capacity of copying records in electronic form			
PAST INSPECTIONS	YES	NO	N/A
PURPOSE OF THIS SECTION: • Determine if there were any findings during past regulatory inspections			
1. When was the laboratory last inspected for GLP compliance by a regulatory body? Which agency conducted the review?			
2. Was the inspection routine or a follow-up from previous inspections with adverse findings?			
3. Obtain and review copies of the most recent inspection reports Were there any adverse findings? For FDA, was a warning letter (Form FDA 483) issued?			

Chapter 3

Study Design

Amy Babb BS*, Jeffrey Ambroso PhD, DABT†, Kelly Davis DVM*, James Greenhaw BS, LAT* and William F. Salminen PhD, DABT, PMP**

*National Center for Toxicological Research, FDA, Jefferson, AR, †Department of Safety Assessment, Glaxo Smith Kline, Research Triangle Park, NC, **PAREXEL International, Sarasota, FL*

Key Points

- Many different study design factors need to be considered for running a successful study
- Some scientific/intellectual ideas may conflict with the logistics of running the study
- The protocol and procedures should be thoroughly reviewed and understood by all parties to ensure that the study will meet scientific, regulatory, and animal welfare needs and can also be conducted smoothly

---

This chapter covers study design and the subsequent chapter covers animal welfare issues for nonclinical studies. Although these are presented in separate chapters, both of these aspects of nonclinical studies strongly influence each other or are intimately intertwined. Therefore, the items raised in both chapters should be considered in designing a nonclinical study. This chapter does not cover detailed study designs related to answering the scientific question at hand. For example, the design of a multigenerational reproduction study or a rat carcinogenicity study is not reviewed. Instead, general study design issues related to the day-to-day running of most nonclinical studies is provided such as animal feeding and care, lighting cycles, randomization to groups, among others. These are all essential to a well-run study that generates scientifically sound data. A checklist is provided at the end of the chapter to help the Contracting Scientist ensure that their study meets the required study design requirements. This checklist can be used in conjunction with the checklist provided in the subsequent animal welfare chapter to ensure that all study design and animal welfare requirements are being met.

There are a variety of guidance documents and books on designing nonclinical studies to meet the scientific question. The following references provide

Nonclinical Study Contracting and Monitoring. http://dx.doi.org/10.1016/B978-0-12-397829-5.00003-X
2013, Published by Elsevier Inc.

detailed information on designing nonclinical and toxicology studies to meet various regulatory and scientific questions:

- International Conference on Harmonisation (ICH)
    http://www.ich.org
- US Food and Drug Administration (FDA) Red Book
    http://www.fda.gov/Food/GuidanceComplianceRegulatoryInformation/
    GuidanceDocuments/FoodIngredientsandPackaging/Redbook/default.
    htm
- Organisation for Economic Cooperation and Development (OECD)
    http://www.oecd.org/document/12/0,3746,en_2649_37465_
    48704140_1_1_1_37465,00.html
- US Environmental Protection Agency (EPA)
    http://www.epa.gov/ocspp/pubs/frs/home/guidelin.htm

Although these references provide very good information on specific study design aspects, they do not cover the more basic issues of designing a study so that it can be practically conducted at the laboratory and meet the necessary animal welfare requirements. Nonclinical studies have both scientific and practical aspects and it is important that the Contracting Scientist understands how the two are related and that not all great scientific ideas can be practically implemented.

When a nonclinical study is contracted to a Contract Research Organization (CRO), especially a Good Laboratory Practice (GLP)-compliant study, the primary focal points are answering the scientific/regulatory question and complying with GLPs. It is easy to lose sight of the fact that there are basic needs of the study that have to be met in order successfully to generate useful data and maintain the health and welfare of the animals. CROs are often put in a difficult situation trying to balance the day-to-day running of a study, complying with animal welfare and GLP requirements, and satisfying the Contracting Scientist's data requirements. Since generating useful data and complying with GLPs is the primary focus of the Contracting Scientist, this is often the focus of the CRO too. However, the CRO can be put in a difficult situation and/or compromise the integrity of the study if some of the Contracting Scientist's needs conflict with the logistical conduct of the study and/or animal welfare requirements. Although the GLPs address some basic animal welfare issues, there are many more regulatory requirements that the CRO must meet that the Contracting Scientist is often not aware of. This and the subsequent chapter will help inform the Contracting Scientist about these requirements so that they can work in a cooperative manner with the CRO to design a study that meets the requirements of all parties. This will also help the Contracting Scientist understand why the CRO might push for certain changes in the study design and why the CRO's Institutional Animal Care and Use Committee (IACUC) and Attending Veterinarian get involved with the study at various points and can be even forceful at times.

GENERAL STUDY DESIGN ISSUES

This section deals with general study design issues that apply to most non-clinical studies. This information will not only help the Contracting Scientist understand the various aspects of getting a study up and running but help identify areas that might pose a conflict between the scientific and regulatory needs and the ability of the laboratory successfully to run the study and generate sound data.

Animal Procurement and Selection

The animal model chosen along with the strain will depend on a variety of factors including but not limited to:

- Regulatory requirements. For example, International Conference on Harmonisation of Technical Requirements for Registration of Pharmaceuticals for Human Use (ICH) guidance for drug development indicates that general toxicity and embryo-fetal development should be assessed in one rodent and one non-rodent species.
- Amount of test material. If the amount of test material is limited, only smaller species may be able to be used until sufficient test material can be manufactured.
- Experience of the laboratory. Certain models require special husbandry and handling techniques and the laboratory should have sufficient experience to conduct these successfully. In addition, historical control data can play an important role in interpreting findings on studies, so the laboratory should have sufficient information from other studies in that model (e.g. ranges for clinical pathology parameters, incidence of developmental abnormalities or various tumors).
- Similarity between the animal model and humans. If you have to choose between multiple species, it is often best to choose the model that is most similar to humans in terms of metabolism and sensitivity to organ toxicity, unless it is a species-specific response that is not applicable to humans.
- Availability. The availability of some models may be limited due to the uniqueness of the model, difficultly breeding, or high demand.
- Research-bred animals. For most studies, it is best to use animals that are raised specifically for research and come with documentation about their health, breeding, and genetics. These animals cost more but you are guaranteed of having high quality animals that are not only healthy but are more likely to provide consistent responses to the test article.
- Animal temperament. For larger animals, temperament can play a major role in successfully running a study. Some breeders either have breeds that are inherently calm and amenable to study procedures or they acclimate the animals to study procedures.

- Inbred versus outbred strains. There may be specific reasons to use an inbred versus outbred strain such as comparing to past studies or obtaining a more consistent response to the test article.

The age of the animals is often dictated by regulatory requirements. It is important to ensure that the animals are of the required age on the day of first dosing. For example, if the animals have to be 6 weeks or younger, animals should arrive at the laboratory when they are 4–5 weeks of age so that they can complete an appropriate quarantine and acclimation period, typically at least a week, after they are received by the laboratory.

The number of animals needed for the study is driven by the study design. If the study incorporates satellite or recovery groups, additional animals will be needed for these groups. It is important to order extra animals for any study to ensure that you only enroll healthy animals. Health issues can arise even when ordering from high quality breeders and sometimes animals need to be excluded from the study. If you did not order extra animals, there would not be a sufficient number of animals to meet the study requirements.

Quarantine and Pre-Study Health Assessment

Most CROs purchase their animals from commercial breeders or they have dedicated facilities for breeding animals. Often, the site of study conduct is a significant distance from the location of animal breeding; therefore, the animals need to be shipped to the study site. Precautions are taken by the breeders and shippers to ensure the health of the animals during shipment; however, the animals are exposed to a foreign environment that can affect their homeostasis and general health. Therefore, after animals are received at the laboratory, they should be quarantined, which typically involves placing them in a dedicated animal room separate from other animals and studies. Shortly after arrival at the laboratory, the health of each animal should be verified by a veterinarian. Sick or diseased animals should be removed from the animal room and isolated to prevent the spread of any potential disease. If the whole population is sick, the animals can remain in the room and treatment administered. However, disease treatments, even when administered prior to the study, could potentially interfere with the study and strong consideration should be given to ordering new animals if disease treatment is required. There is no hard and fast rule about separating the quarantine and acclimation periods or wrapping them into a single period. The key is to ensure that the animals are sufficiently recovered from any shipping-related stress, are acclimated to the laboratory environment, and only healthy animals are enrolled. The latter part can be verified by a veterinarian sometime during the acclimation period and this is often separate from the health check conducted after the animals arrive on site.

Most studies use completely naïve animals. As mentioned above, pre-study treatments may alter the response of the animals to the test material, for example, by inducing metabolic enzymes. If pre-study treatments are used to eradicate a disease or prevent a disease (e.g. vaccinations), they should be separated from test article administration by sufficient time so as to have the least possible influence on the response to the test article. If pre-study treatments are required, consideration should be given to extending the acclimation period so that the animals are fully recovered and healthy. Some studies may use animals that have been used on previous studies (e.g. some nonhuman primate studies). For these studies, a sufficient wash out period must be used between studies to ensure that there are no residual effects of the first test article on the response to the second test article. In addition, the health of the animals should be thoroughly evaluated prior to placement on a new study.

In order to ensure that the animals remain healthy and disease free during the study, sentinel animals are often included in the animal room, especially for longer-term rodent studies. These animals are checked for the presence of various diseases before, during, and at the end of the study.

Animal Identification

Animals need to be appropriately identified. A variety of methods are available including tattoos, ear tags, radio frequency identification (RFID) chips, collars with tags, ear punches, and toe clips. The latter two methods are less preferable from an animal welfare perspective and should not be used unless there are no other feasible means. The method used should enable easy and unambiguous identification of each individual animal. In addition, the animal identification information should be provided on the outside of the housing unit. When performing study functions, animal identification must be constantly verified to ensure the right animal is in the right housing unit and is receiving the right treatment. RFID chips are very efficient since the animal identification is verified by scanning the animal. If the wrong animal is selected, the computer will give an error and/or indicate the actual treatment that needs to be administered. If the doses are also bar-coded and have to be scanned and linked to the relevant animal/dose group, this provides another layer of verification that the right dose is being administered to the right animal. If computer verification methods are not used, a simple method is visually verifying the animal identification each time the animal is handled for any study function (e.g. if rat number 21 is selected for dosing, when the rat is removed from the cage, the technician verifies that the tail tattoo is 21).

Animal Housing

There are pros and cons to different types of housing. For rodents and smaller mammals, the typical choices are wire bottom cages or plastic cages using contact bedding. Wire bottom cages can lead to sores on the feet but they enhance cleanliness since urine and feces fall through the bottom of the cages. Plastic cages utilize contact bedding and are generally preferable from an animal welfare perspective. When selecting the contact bedding, it is important to ensure that it does not interfere with the study since some types of bedding contain compounds that can alter the metabolism of drugs (e.g. pine wood chip bedding is known to induce hepatic drug metabolizing enzymes). For rabbits, cages are typically used and they need to be cleaned/changed frequently due to the caustic urine. Cats are typically housed in cages with litter boxes that need changing. For dogs, cages or runs are typically used. For both cages and runs, raised flooring that has holes facilitates cleaning but can lead to sores on the feet. Runs that use contact bedding are labor intensive but can be preferable from an animal welfare perspective. Nonhuman primates are often housed in cages or larger pens.

There are also pros and cons to group versus individual housing. Some study designs preclude group housing if they require individual clinical observations and food and water consumption. In these situations, individual values could not be determined if group housing was used since findings such as diarrhea would have to be assigned to all animals and consumption values would be for all animals in the housing unit. The advantage of group housing is that it allows social interactions between animals and is preferable from an animal welfare perspective; however, precautions must be taken to ensure that co-housed animals are compatible (i.e. they do not fight). In addition, co-housing can lead to confounding of study findings such as might occur with copraphagia during a pharmacokinetic study that leads to secondary exposures to the drug and/or metabolites via fecal consumption.

The size and construction of the housing varies by species and needs of the study. Animal welfare regulations dictate minimum size requirements and provide some very basic criteria for construction (e.g. easily cleanable and sanitizable surfaces and no sharp edges). For dogs, housing them in larger runs can preclude the need for exercise periods, which may be beneficial for studies that need to maintain separation of the animals. When animals are group housed, the minimum size of the caging increases; however, the area required per animal decreases since the increased socialization is taken into consideration.

Environmental enrichment should be used to enhance the well-being of animals. A variety of different environmental enrichment approaches can be used and these vary by the species. Regardless of the approach, it should be determined if the enrichment will interfere with the scientific purposes of the study or health of the animals. The ultimate goal is to provide enrichment that is compatible with the scientific purpose of the study and maximizes animal health and well-being.

Water and Feed

Water must be available at all times unless specific withholding periods are required by the study. Water can be supplied via various means such as bottles or automated watering lines. When automated watering lines are used, it is important to ensure that the animals are acclimated to the system (i.e. they learn how to use them). For all water systems, they should be frequently cleaned and sanitized. Water samples should be collected at various points throughout the facility and tested for contaminants (e.g. microorganisms, pesticides, and heavy metals) that might interfere with the health of the animals.

Wholesome feed must be available to the animals, preferably certified feed that has been tested by the manufacturer to ensure that it meets specific nutritional requirements and lacks various contaminants (e.g. microorganisms, pesticides, and heavy metals). If possible, the same lot of feed should be used for the duration of the study or the number of batches should be limited (e.g. on chronic rodent studies). The duration of feeding can vary by species and study. Rodents, rabbits, and cats are typically fed *ad libitum*. Dogs can be fed *ad libitum*; however, they often play with and spill their feed. Therefore, a short duration of feed offering each day can be beneficial since the dogs are more likely to consume their feed instead of playing with it. Nonhuman primate diets are often supplemented with fresh items, such as fruits. Some studies require a fasting period before some study functions (e.g. clinical pathology blood draws, necropsy) or dosing. It is important to adhere to the specified fasting time windows to provide the most consistent response possible.

Environmental Controls

The environment of the animal room must be controlled to minimize any influence that environmental changes might have on the response of the animals to the test article. The animal environment can be divided into the macroenvironment, which is the whole animal room, and the microenvironment, which is the primary housing unit of the individual animal (e.g. cage or run). Normally, the macroenvironment is tightly controlled within animal- and protocol-specific specifications. The microenvironment is more difficult to control unless specialized caging is utilized, such as microisolator units for rodents that provide cage-by-cage airflow.

The animal room should be maintained at a relatively constant temperature that is specific to the species, strain, and age of the animals. Humidity should also be controlled, but the acceptable variance for humidity is a lot wider than for temperature. Low humidity can lead to drying of eyes and mucous membranes; whereas, high humidity can promote bacterial and fungal growth and lead to increased odors since waste will not dry as readily. For large animals, humidity will often spike during cleaning operations that utilize water. As long as the spike is for a short duration, it is unlikely to have a major impact on the animals. If continuous environmental monitoring systems are used, this could result in an out-of-range alarm and could be a protocol deviation. Therefore, it can be beneficial to check humidity at specific times of the day that are before or well after the daily cleaning operations.

The light cycle should be controlled to maintain consistent diurnal cycles since the level of a variety of endogenous molecules changes over the course of the day (e.g. some hormone levels). The actual on/off times for the lights might need to be tailored for specific species or study purposes (e.g. to promote breeding) and this would need to be specified in the protocol to avoid deviations. Most animal rooms lack windows and/or skylights and this is preferable from the perspective of maintaining a consistent light/dark cycle. The lack of external windows also enhances security by limiting external access to the room. A small window in the door entering the animal room is usually used to enable easy observation without entering the room and this is acceptable as long as the corridor lights are also on a similar light cycle.

The room should have sufficient ventilation. It is preferable to use only fresh air to supply the animal rooms instead of recirculated air. If recirculated air is used, precautions should be taken to ensure that it is pure (e.g. HEPA filtered). Normally, a specified number of complete room air changes (e.g. 10–15 per hour) are used to ensure adequate ventilation and removal of odors. Diffusers and/or multiple supply vents should be used to ensure that there is uniform air delivery throughout the room. Most studies will utilize a positive pressure system where the animal room is at a higher pressure than the outside hallway. This prevents contaminants from entering the room from the corridor that might interfere with the study. However, some studies, especially those using infectious or highly hazardous agents, might utilize a negative pressure setup to ensure that the agent or test article remains within the room. Also, studies using a clean/dirty corridor setup might have the room being negative pressure relative to the clean corridor and positive pressure relative to the dirty corridor.

As mentioned above, the animal microenvironment is often different than the room macroenvironment. The microenvironment will vary depending on the type of caging that is used and the position of the caging within the animal room. For example, rats housed on the bottom of the racks will receive less light and might be exposed to lower temperatures than those on the top racks. Depending on where the airflow originates within the room, some animals might have more airflow within their cages resulting in less waste odor and byproducts. In order to minimize the influence of cage position, it is important to ensure that each dose group is equally represented at various rack positions. Alternative and/or additional approaches are to rotate cage positions on the rack or within the room at a specified frequency or to randomize cage positions on the rack or within the room.

Randomization to Groups

Before dosing begins, the animals should be randomized to treatment groups. The first step is to cull unhealthy animals and animals that are outside of the protocol-specified weight range, if applicable. There are various ways to randomize the animals to dose groups; however, the main goal is to randomize the animals to dose groups while maintaining a similar weight distribution for each group. Weight is chosen as a common variable since it can influence a variety of different endpoints and it is best to ensure similar weights between groups. The simplest method is randomly to assign numbers to each animal and assign them to groups in the order of the numbers. The mean weights and weight distribution for each group is compared and if outside of protocol speci-fications, the randomization is repeated. Another method is to rank the animals from lowest body weight to highest body weight. The animals are then blocked in groups corresponding to the number of dose groups (e.g. if there are control, low, and high dose groups, the animals are blocked in groups of three animals). Random numbers are then generated corresponding to the different dose groups (in the example above, the random numbers would be 1, 2, and 3 generated in a random order) and assigned sequentially to each animal in the block. Once the animals are assigned to the dose groups, the similarity of the mean weights and weight distributions is verified and the randomization can be repeated if outside of protocol specifications. Since males and females vary significantly in their weights, separate randomizations should be conducted for the two sexes.

Dosing

Dosing can be done by various routes of exposure and this is typically driven by the study protocol. Dosing should be done at the same time of day and for the same duration, when applicable, to minimize the influence of any diurnal effects. For oral exposure, the test article may be a liquid or solid. For liquids, oral gavage, especially in rodents, is routinely used to ensure that all of the test article reaches the stomach. For larger animals, liquid formulations may be able to be administered within the mouth followed by swallowing; however, it is important to ensure that the animals truly consume the whole dose and do not let any escape from the mouth. For solid formulations (e.g. pills or capsules) with larger animals, they may be administered using a pill gun or by finger manipulation. Dermal exposures should use some method to prevent oral inges-tion of the test article (e.g. dressing or collar or placement on the back in an area that cannot be reached by the animal). In addition, for dermal exposures, if the test article is a liquid or lotion applied to an unprotected area of skin, consider-ation should be given to animal-to-animal transfer via inter-animal contact (if group housed or caging design allows animal to animal contact) and via animal handlers (e.g. by transfer via the gloves). Inhalation exposure can involve nose only or whole body exposure. For whole body exposure, oral ingestion is likely during grooming and this should be considered. Nose only exposure involves restraint of the animals; therefore, the exposure duration should be kept as short as possible to minimize animal stress. Other routes of administration have other unique requirements (e.g. intravenous administrations may be bolus or pro-longed infusions and require a technician skilled in this route of administration).

Dosing can be stressful to the animals and every means should be taken to minimize the stress. For example, routine handling of animals from a young age often prepares them for more stressful procedures, such as oral gavage dosing. Some animals undergo sham dosing procedures from a young age so that they are acclimated to the dosing procedures once the study starts. Dosing should never be forced since not only does that increase animal stress but it can lead to dosing errors and even animal death (e.g. if some of the dose solution is accidentally administered within the lung during oral gavage dosing). It is important that the technician conducting the dosing is thoroughly trained and experienced with the dosing procedures for the given species since an inexperienced technician is more likely to cause animal stress and have dosing errors due to difficulty in administering the dose.

One confusing point about the start of dosing is how different laboratories refer to the first day of dosing. Some will designate the first day of dosing as Study Day 0 or Study Day 1. Yet others may designate the first day of dosing relative to the first day of data collection (e.g. Study Day 3 if acclimation body weights were collected on Study Day 1). Regardless of the designation, it is important to understand the laboratory's specific designations and this should be included in the study protocol.

One last point about dosing is that some studies require a staggered start to accommodate subsequent procedures, such as necropsies. It is important to understand exactly how the study will be conducted so that you can ensure that the staggered dosing does not interfere with the requirements of the study.

In-Life Evaluations

There are many different types of clinical evaluations that can be conducted during the in-life phase of the study, which starts once animals are enrolled in the study and in-life data collection begins until data are no longer collected on live animals (e.g. once a necropsy is conducted on the animals). The start of data collection could be during the acclimation phase if pre-dose baseline assessments are needed or it could be after the first dose is conducted. Most nonclinical studies use a standard battery of in-life evaluations since they provide insight into the response of the animal to the test article. Some specialized study designs (e.g. carcinogenicity or reproduction studies) may lack some of these standard evaluations and include additional specialized evaluations.

A detailed physical exam should be conducted on each animal during the acclimation period to ensure that only healthy animals are enrolled in the study. This should be ideally conducted by an experienced veterinarian. A physical exam may also be indicated during the study, particularly for large animals, to assess thoroughly the health of the animals by a trained veterinary professional.

During the study, the animals should be routinely observed by the technicians for adverse effects. These can be simple health checks that do not involve removing the animal from the cage to more complex assessments that involve removing the animal from the cage for a thorough health and behavior assessment. At a minimum, daily mortality/moribundity and feed/water checks need to be conducted on each animal. However, once dosing starts, more detailed health assessments should be conducted, typically shortly after the morning dosing and then later in the day. These assessments are used to determine if the test article is affecting the health or behavior of the animals by observing changes in the skin, fur, eyes, mucous membranes, respiratory and somatomotor function, circulatory, autonomic and central nervous systems, general behavior, and any other abnormal findings.

Body weight is recorded for most studies and changes in body weight or the rate of body weight gain is a frequent effect of many test articles. Body weight can be recorded daily, weekly, or at some other interval as long as an adequate picture of the changes over time can be determined. It may also be beneficial to record body weight during the acclimation period so that individual animal changes can be more accurately assessed once dosing starts (e.g. under untreated conditions, some animals may not gain weight as quickly as others).

Feed consumption is also recorded for most studies. Feed consumption is often determined on a daily or weekly basis. Since feed wastage is common (e.g. dogs commonly spill feed outside of their housing unit and mice and rats likely spill some feed into their cage during consumption), feed consumption is often more variable than body weight, but it still provides insight into possible effects of the test article.

Water consumption can be measured; however, this entails using a container that facilitates recording water consumption (e.g. a water bottle). Therefore, automated watering systems cannot be used if water consumption data are required. In addition, water consumption is typically highly variable due to spillage during consumption and/or leaking containers.

Ophthalmology exams are conducted in many nonclinical studies. They are often conducted during the acclimation period to ensure the eyes are normal prior to test article exposure. It is best if the eye exams are conducted by a qualified veterinarian and even better if conducted by a board certified veterinary ophthalmologist. Eye exams are then conducted at various times during the study and at least shortly before the necropsy, when applicable.

In large animal studies (e.g. dogs), cardiovascular assessments, including electrocardiograms (ECGs), are often conducted. They are typically conducted during the acclimation period to establish a baseline for each animal and then during the study. It is best if the examinations are conducted by a qualified veterinarian and even better if conducted by a board certified veterinary cardiologist. The ECGs can be recorded by a qualified technician or veterinarian; however, they should be read by a board certified veterinary cardiologist. Additional cardiovascular examinations and ECGs are conducted at various times during the study and at least shortly before the necropsy, when applicable.

Clinical pathology assessments are conducted in many nonclinical studies. Clinical pathology typically consists of a battery of hematology (e.g. red and white blood cell counts and differential) and clinical chemistry (e.g. liver enzymes, electrolytes, protein) tests. Blood volume for these tests is typically not an issue for rats and larger animals but can be a restriction in mice. Specially sized small collection tubes are very useful for mice since they allow very small volumes to be collected. Since some clinical chemistry tests take a larger volume of serum/plasma, it may be necessary to drop some of the individual tests in order to accommodate the small sample volume. The collection tubes that are needed should be verified with the clinical pathology laboratory since some laboratories prefer certain types of specimens (e.g. serum instead of plasma for clinical chemistry). It is also important to consider how long the samples can be stored prior to analysis and if the blood collection timepoints allow the samples to be analyzed by the clinical pathology laboratory in a timely manner (e.g. will the samples sit overnight after collection or will they be analyzed immediately). Sometimes baseline clinical pathology samples will be collected during the acclimation period. Subsequent samples are then collected at various times during the study and then typically at necropsy, when applicable. Depending on the study design, blood volume withdrawal limits may pose an issue for rats, especially if the baseline draw is close to the first post-dose draw and/or additional tests requiring blood are being conducted. Typical blood volume withdrawal limits for various species can be found in Diehl et al. 2001[1]. Although these are recommendations, most laboratories will need adequate scientific justification for withdrawing larger volumes. While the animals are alive, blood can be withdrawn from any accessible vein. The retro-orbital sinus can also be used for rats and mice, which involves using a capillary tube to pierce the sinus behind an eye and wick up the blood. Since blood volume is limited in mice, in-life blood draws can be problematic. For terminal collections, cardiac puncture is often used for mice and rats that have been euthanized with carbon dioxide. Blood is typically collected from larger animals via an accessible vein shortly before euthanasia. The order of blood collection and analysis should be determined. Some studies may require blood collection from all animals in one group followed by the next; whereas, others may require randomized collections among the groups and randomized sample analysis or some other collection scheme to minimize sample collection bias.

Urine is sometimes collected for urinalysis. However, it can be difficult collecting urine that is not contaminated by other substances such as feces and spilled feed and water unless a direct collection method is used such as cystocentesis. Special caging may help collect cleaner urine such as metabolism cages for rodents. Another problem with urine collection is bacterial growth since the urine is typically collected over an extended period (e.g. overnight). Surrounding the urine collection container with ice can help but may not be feasible with certain cage designs (e.g. if urine is collected from cage pans or troughs).

Terminal Procedures

Necropsies are conducted on many nonclinical studies so that effects of the test article on various organs, both macro- and microscopically, can be assessed in detail. For many toxicology studies, histopathological changes in organs that are observed by light microscopy are often the effects that determine the lowest dose of

---

[1]Diehl, K.H., Hull, R., Morton, D., Pfister, R., Rabemampianina, Y., Smith, D., Vidal, J.M., & vande-Vorstenbosch, C. (2001). A good practice guide to the administration of substances and removal of blood, including routes and volumes. *Journal of Applied Toxicology, 21,* 15–23.

the test article that caused toxicity. When planning for the necropsy, it is important to consider if it needs to be tightly timed relative to the dose (e.g. 6 hours ± 15 minutes) and if any additional sample collections are required beyond what is normally collected for a nonclinical study. Also, the protocol should clearly outline the procedure that should be used for unplanned animal deaths or moribund animals (e.g. should dead carcasses be refrigerated and sent to necropsy or simply be discarded; should moribund animals be euthanized and subject to a complete necropsy).

A necropsy can be relatively simple if just a single organ needs to be collected and preserved for histopathology or extremely complex and time consuming if every major organ needs to be collected and some of them have to be preserved in a manner that is compatible with analysis for gene expression or other sensitive analyses. Regardless of the complexity, it is important to determine the pathology group's ability to conduct the necropsies and meet all of the study requirements. For example, how many animals can be necropsied per hour, can additional technicians be used to increase throughput, or are there sufficient necropsy stations to accommodate the study. Ideally, a board certified veterinary pathologist will either be present at the necropsy or at least available for immediate consultation whenever the necropsy technicians have questions (e.g. if a certain finding on the animal should be recorded and collected as a gross lesion). If tightly timed necropsies are needed, this will require close coordination between the group conducting the dosing and the pathology group. Also, consideration needs to be given about the order of the necropsy. In order to minimize diurnal influences, especially for terminal blood collections and necropsies that involve many animals over many hours, it may be best to conduct the necropsy on an animal from group 1 then an animal from group 2 and then an animal from group 3 for a three group study. The necropsy then cycles back to the next animal in group 1 then group 2 and so on until all the animals are completed. A staggered start study may also be needed to accommodate a large number of animals where necropsies are conducted over several days.

Animals are often fasted overnight prior to the necropsy unless the study specifically requires continuous feeding. This helps eliminate waste from the gastrointestinal tract and can improve the appearance of some tissues (e.g. glycogen stores in the liver are reduced leading to hepatocytes with less vacuolization). Immediately prior to the necropsy, blood may be collected. The animal will be euthanized, and the necropsy will be conducted. The last clinical observations should be available at necropsy. For a typical toxicology study, the list of organs is extensive and involves the collection of over 50 different samples. Organ weights for select key organs are typically measured and any gross lesions that are observed are recorded. It is important that the gross lesions are noted, collected, and preserved so that the pathologist can correlate the gross findings with histopathological changes. The organs are then preserved in 10% neutral buffered formalin but some tissues such as the eyes and testes might be preserved in special fixatives in order to maintain their morphology, such as Modified Davidson's fixative. When desired, methods may be used to distinguish certain paired organs (e.g. left versus right kidney). Depending upon the species and/or the size of a particular organ, a whole organ may be placed in fixative

STUDY DESIGN	YES	NO	N/A
Animal Procurement and Selection			
1. Animals are obtained from a commercial breeder			
2. Animals are bred specifically for research			
3. Animals have appropriate documentation Breeding and/or genetics records Medical records for larger animals Regulatory records (e.g. USDA number)			
4. Animals meet the designated regulatory requirements of the study (e.g. correct species, strain, sex, age)			
5. A sufficient number of animals can be obtained for the study (i.e. animal supply is not an issue)			
6. Animals have a calm temperament and/or the breeder habituates animals to mock study procedures			
7. The laboratory has experience with the species and strain of animal			
8. Sufficient historical control data are available for the selected species and strain of animal			
9. The selected species and strain is the best for modeling human exposures (e.g. similar pharmacokinetics, most sensitive species, similar target organ toxicity), when applicable			
10. Sufficient test material is available to complete the study in the selected species			
11. Animals are obtained and acclimated at an age that meets the maximum age limits on the first day of dosing (e.g. animals are ≤7 weeks of age on the first day of dosing)			
12. Extra animals are ordered to ensure only healthy animals are enrolled in the study			
13. Animals are naïve or any pre-study treatments (e.g. dewormers or vaccines) do not interfere with the study			
14. If animals have been used on another study, a sufficient washout period has occurred between studies, a complete health assessment has been conducted, and only healthy animals are enrolled			
Quarantine and Pre-Study Health Assessment			
1. Shipping precautions are taken to minimize animal stress			
2. Animals are quarantined when they arrive at the laboratory How long is the quarantine period? Where are the animals quarantined? Who releases animals from quarantine and how is this decision made?			
3. Animal health is assessed by a veterinarian after arrival How soon after arrival is the animal's health assessed? What assessments are conducted?			
4. Animals undergo an acclimation period Is the acclimation period separate from the quarantine period? How long is the acclimation period?			
5. A complete physical and health assessment is conducted by a veterinarian on each animal during the acclimation period to ensure that only healthy animals are enrolled in the study			

STUDY DESIGN	YES	NO	N/A
6. Sick or diseased animals are removed from the room unless the entire room is sick or diseased			
7. If sick or diseased animals are treated and then enrolled in the study, the treatments do not interfere with the study			
8. Are sentinel animals being used on the study to ensure that the animals on study are not exposed to any diseases? How many are being used? How often are they tested? Is the battery of tests appropriate for the species?			
Animal Identification			
1. Animals are appropriately identified How is each animal identified?			
2. Ear punches and toe clips for rodents are only used when no other feasible method is available			
3. Each primary enclosure (e.g. cage) is labeled with the animal identification			
4. Animal identification is verified each time an animal is removed from or placed back in a primary enclosure What is the verification procedure?			
5. Animal identification is verified prior to dosing, observations, etc. What is the verification procedure?			
Animal Housing			
1. What type of primary enclosure is used (e.g. cages, runs)?			
2. If wire-bottom cages are used for rodents and rabbits, is their use scientifically justified?			
3. If wire-bottom cages are used for rodents and rabbits, has consideration been given to prevention of foot lesions?			
4. If raised flooring with holes (e.g. holes, slats, grates) is used for larger animals, is it constructed to minimize foot lesions?			
5. Is the housing constructed to minimize animal injury (e.g. no sharp edges, entrapment risk)?			
6. Is contact bedding used? Does it interfere with the study (e.g. pinewood chips induce liver metabolism)?			
7. Is environmental enrichment used? Does it interfere with the study or health of the animals?			
8. Is group housing used? Are the animals compatible and who makes this determination? Does group housing interfere with the scientific purposes of the study (e.g. individual feed consumption and clinical observations are needed)?			
9. Does the size of the primary enclosure meet the regulatory and study requirements (e.g. large runs for dogs may be needed if they can not be exercised together)?			

STUDY DESIGN	YES	NO	N/A

3. Detailed clinical observations are conducted
   When do they start relative to the first day of dosing?
   How often and at what time of day are they conducted?
   Are the animals removed from the cage for the observations?
   What are the observations being conducted and do they meet the study requirements?			

4. Body weight is recorded
   How frequently?
   Who records the terminal body weight (e.g. in-life group or pathology)?			

5. Feed consumption is recorded How frequently?			

6. Water consumption is recorded
   How frequently?
   How is it measured (e.g. weighing water bottles)?			

7. Eye exams are conducted during the acclimation period
   Are eye exams conducted at other times during the study?
   Who conducts the eye exams (e.g. staff veterinarian)?			

8. Blood and/or urine for clinical pathology are collected Which tests are being run (e.g. hematology, clinical chemistry, urinalysis)?			

9. What collection tubes are needed for the blood (e.g. EDTA, serum separator)?
   What size tubes are needed?
   How much blood will be collected?
   Can this amount be successfully collected from the species?			

10. When will blood and/or urine be collected (e.g. in-life and/or terminal)?			
11. What is the route of collection (e.g. retro-orbital sinus, vein, cardiac puncture)?			
12. How long will the samples be stored before analysis?			
13. Does the blood collection need to be completed within a certain time of day to prevent diurnal influences (e.g. hormone level changes)?			
14. Does the blood need to be collected and/or analyzed in a certain order or randomized (e.g. by group or in random order)?			

15. Is urine collected for urinalysis? Are methods used to limit:
   Feed, water, and fecal contamination
   Bacterial growth			

16. What additional in-life evaluations are being conducted?			
17. When do these evaluations start (e.g. during acclimation, after 1st dose)?			
18. How frequently are these evaluations conducted (e.g. daily, weekly)?			
Terminal Procedures			

1. The protocol specifies whether or not the following animals should be subject to a necropsy:
   Animals found dead
   Moribund animals			

STUDY DESIGN	YES	NO	N/A
2. How many necropsies can be conducted per day? Does this meet the study requirements?			
3. Do the animals need to be fasted overnight before the necropsy?			
4. Necropsy technicians are trained and experienced with the species			
5. Technician drawing terminal bloods is experienced and can withdraw the needed blood volume consistently			
6. Gross necropsy is being conducted Entire GI tract is opened and observed?			
7. Selected organs are being weighed			
8. Tissues are being collected for histopathology Fixation method? Lungs and bladder are infiltrated with fixative? Any special fixatives used for select tissues (e.g. eyes, testes)?			
9. Gross lesions are documented, collected, and preserved			
10. Tightly timed necropsies are required (e.g. 6 h ± 15 min)? Has this been coordinated with the group conducting dosing? Does the pathology group have sufficient manpower and stations? Is the estimate for the amount of time per necropsy accurate and/or has a test necropsy on an extra animal been conducted?			
11. Are special collections required (e.g. tissue quickly preserved for gene expression analysis)?			
12. A veterinary pathologist will supervise the necropsy			
13. What is the order of necropsy (e.g. one animal per group and cycling back to the first group)? Is the time between dosing and necropsy similar for each animal? Does this minimize diurnal effects?			
14. Pathologist reviews in-life data (e.g. clinical observations, clinical pathology, bodyweight) before reading slides			
15. Pathologist needs to be "blinded" to treatment while reading slides			
16. Only control versus high dose group slides are read by pathologist to start with			

Chapter 4

Animal Welfare

Neera Gopee DVM, PhD, DABT*, Jeff Carraway DVM, MS, DACLAM*,
Lady Ashley Groves BS, BA* and William F. Salminen PhD, DABT, PMP†

*National Center for Toxicological Research, FDA, Jefferson, AR, †PAREXEL International, Sarasota, FL

> **Key Points**
> - Ensuring animal welfare is not only a regulatory requirement but it helps to generate scientifically sound data
> - The Institutional Official, Attending Veterinarian, and Institutional Animal Care and Use Committee play key roles in ensuring animal welfare at the laboratory
> - The Contracting Scientist needs to be aware of how animal welfare requirements can impact the design and conduct of his/her study

Before the 1960s, animal welfare in scientific research was essentially left up to individual researchers. While some studies incorporated adequate animal welfare provisions, others incorporated procedures and husbandry practices that entailed unnecessary animal suffering. Due to basic animal welfare concerns and in order to ensure useful data generation (i.e. studies were not confounded by the use of sick or stressed animals), the first US guideline for research animal care was published in 1963, titled: *The Guide for the Care and Use of Laboratory Animals.* These voluntary guidelines were developed from the input of veterinarians and scientists who understood that successful and reliable studies depended on consistent, high-quality animal care. Shortly afterwards, in 1966, the Laboratory Animal Welfare Act (AWA) (Public Law 89-544 Act) was passed. The AWA was the first major federal regulation that mandated specific requirements for animals used in research and has been amended several times since.

Meeting animal welfare requirements is important for regulatory, scientific, and ethical reasons. Scientific research on animals is supported by many people because it is the only reliable mechanism for testing the safety and efficacy of compounds before exposure to humans. However, most people also support minimizing animal suffering and ensuring the ethical treatment of animals used in research. The various mandated and voluntary animal welfare regulations

Nonclinical Study Contracting and Monitoring. http://dx.doi.org/10.1016/B978-0-12-397829-5.00004-1
2013, Published by Elsevier Inc.

aim to balance these needs by allowing animal research to be conducted under conditions that minimize animal suffering whenever possible. When a study complies with these animal welfare regulations and policies it can be defended as being conducted under conditions that minimized animal pain and distress.

Some basic animal welfare requirements were covered in the previous chapters. This chapter will provide an overview of general animal welfare requirements and the regulations and standards, both mandated and voluntary, which drive them. The focus will be on *The Guide for the Care and Use of Laboratory Animals* ("Guide"), 8th edition (National Research Council of the National Academies, The National Academies Press, Washington, DC, 2011) since the standards presented in the "Guide" are internationally accepted as state-of-the-art animal welfare practices. It is beyond the scope of this chapter to provide detailed information on every animal welfare regulation in the USA and throughout the world since there may be unique state, local, and regional requirements. Instead, the salient features of the "Guide" and the components that make up a successful animal care program will be reviewed. This information will help the Contracting Scientist understand the various requirements that must be met and ones that are not required but should be met whenever possible.

In the USA, there is a complex series of regulations and guidelines covering animal welfare that vary depending on the species being used and funding mechanism for the study. The AWA is one of the main regulations covering research on animals. The AWA is enforced by the United States Department of Agriculture (USDA) Animal, Plant, and Health Inspection Service (APHIS). As mandated by the AWA, the USDA developed standards governing the humane handling, care, treatment, and transportation of animals used in research, which are listed in the Code of Federal Regulations, Title 9, Chapter 1, Subchapter A, Parts 1, 2, and 3. These standards include minimum requirements for the handling, housing, feeding, watering, sanitation, ventilation, shelter, separation of species, and adequate veterinary care. The AWA covers all species (live and dead) used for research, testing, teaching, experimentation or exhibition except mice of the genus *Mus*, rats of the genus *Rattus*, birds, horses and farm animals for use as food- or fiber-related activities. Under these regulations, each facility using a covered species must register with the USDA and comply with the specific requirements. Each facility, with the exception of US federal agencies, is subject to an unannounced yearly inspection by the Animal Care section of the USDA's APHIS to ensure compliance with the regulations.

In the USA, in addition to the AWA, research using mice, rats, and other vertebrate animals is covered by the Public Health Service (PHS) Policy on Humane Care and Use of Laboratory Animals (PHS Policy) on behalf of the US Department of Health and Human Services (DHHS), but only if the study is conducted or funded by the PHS. The Office for Laboratory Animal Welfare (OLAW) is responsible for enforcing the PHS Policy. Many of the PHS studies are conducted under research grants funded by the National Institutes of Health; therefore, research using mice (of the genus *Mus*) and/or rats (of the genus *Rattus*) that are not funded by the PHS (e.g. pharmaceutical company sponsored research) are not covered by any federal regulations from an animal welfare perspective. In these cases where mandated regulations do not apply, it is important that the study conforms to various voluntary standards (e.g. standards in the "Guide") so that it can be defended as complying with state-of-the-art animal welfare practices. In addition, even for regulated studies, it is best if the study complies with minimum regulatory requirements as well as the voluntary standards since this will ensure the highest level of animal welfare.

Table 4.1 highlights the US federal regulatory jurisdiction for typical mammalian species used in laboratory research. Various state and local regulations may also cover animal welfare and various international regulations cover animal welfare outside of the USA.

Regardless of the global location of the study, there are various voluntary standards and certifications that help ensure animal welfare. The Association for the Assessment and Accreditation of Laboratory Animal Care International (AAALAC) is a private, nonprofit organization that promotes the humane treatment of animals in science. AAALAC accredits organizations that meet or exceed applicable animal welfare standards. AAALAC accreditation is voluntary and involves peer review of the facility by qualified animal welfare experts. Continued accreditation involves inspections of the facilities and animal care program at least every three years. The three primary standards that AAALAC uses in determining the level of animal welfare are: (1) The "Guide"; (2) the

TABLE 4.1 US Federal Animal Welfare Regulations for Mammalian Species

Funding	Species	Mandated regulations	Voluntary standards
Private	Mice of the genus *Mus* Rats of the genus *Rattus*	None	The "Guide"
	All mammals (live or dead) except mice of the genus *Mus* and rats of the genus *Rattus*, horses, and farm animals for food or fiber	AWA (enforced by USDA)	
PHS	All mammals	PHS Policy (enforced by OLAW)	The "Guide"
	All mammals (live or dead) except mice of the genus *Mus* and rats of the genus *Rattus*, horses, and farm animals for food or fiber	AWA (enforced by USDA)	

AWA: US Animal Welfare Act; OLAW: Office of Laboratory Animal Welfare; PHS: US Public Health Service; The "Guide": *The Guide for the Care and Use of Laboratory Animals*; USDA: US Department of Agriculture.

Guide for the Care and Use of Agricultural Animals in Research and Teaching, FASS 2010; and (3) the European Convention for the Protection of Vertebrate Animals Used for Experimental and Other Scientific Purposes, Council of Europe. AAALAC accreditation can be obtained by any laboratory throughout the world and is a globally recognized certification of a laboratory's compliance with animal welfare standards and dedication to continually improving animal welfare. Whenever possible, studies should be conducted at facilities that have AAALAC accreditation since those facilities are dedicated to a high level of animal welfare. If a facility loses AAALAC accreditation, this is a major concern and could be a sign that not only is animal welfare being compromised but that study integrity may be compromised due to below average animal care and use.

Ensuring day-to-day animal welfare is a key responsibility of the technicians who handle and treat the animals and the veterinarians who oversee the studies and facility. There are several certifications that show a technician's or veterinarian's commitment to laboratory animal welfare. The American Association for Laboratory Animal Science (AALAS) is a nonprofit association of professionals that advances responsible laboratory animal care and use to benefit people and animals. AALAS has three levels of technician certification: (1) Assistant Laboratory Animal Technician (ALAT); (2) Laboratory Animal Technician (LAT); and (3) Laboratory Animal Technologist (LATG). In many laboratories running nonclinical Good Laboratory Practice (GLP) studies, these certifications are a common requirement for increasing levels of laboratory animal care positions. For veterinarians, the American College of Laboratory Medicine (ACLAM) administers a certification program in which successful candidates become Diplomates of the ACLAM. A laboratory's dedication to animal welfare can be assessed by the number of technicians and veterinarians who are certified and the certification requirement prerequisites for holding animal care and use positions.

In the following sections, the salient standards outlined in the "Guide" will be reviewed. In addition, some specific requirements established by the USDA under the AWA will be covered since the AWA covers several unique animal welfare aspects (e.g. reporting requirements) and is the main regulation covering animal welfare in nonclinical studies conducted in the USA. At the end of the chapter, an animal welfare checklist is provided to help the Contracting Scientist audit a laboratory's animal care and use program.

THE "GUIDE"

The "Guide" was developed by the Institute of Laboratory Animal Resources (ILAR), Commission on Life Sciences, National Research Council and encompasses standards representing state-of-the-art animal welfare practices that are internationally recognized. AAALAC uses the "Guide" as one of three key standards for evaluating a facility's animal welfare practices. The "Guide" specifically takes into consideration regulatory requirements of the AWA and PHS Policy since they drive most US-based activities involving animal research. Therefore, the "Guide" plays a pivotal role in ensuring that a facility is using state-of-the-art animal care and uses procedures that ensure animal welfare. Since the "Guide" is globally recognized, all animal studies, regardless of location, should be conducted according to the practices in the "Guide" unless justified otherwise for scientific reasons. Conducting studies that deviate from the "Guide" without sufficient scientific justification needlessly places animal welfare at risk. In addition, these studies are not defensible to the public as being conducted using state-of-the-art animal welfare practices.

The "Guide" provides broad recommendations using both performance and engineering standards for achieving specific outcomes regarding animal welfare but it leaves it up to the individual facilities on how they reach these goals. The "Guide" emphasizes a performance-based approach since many variables (e.g. species, expertise, facility, study design, etc.) make a prescriptive engineering approach impractical for broad-based implementation across varied facilities and study types. Performance standards specify an outcome and provide criteria for assessing that outcome, but provide flexibility and do not limit the methods by which to achieve that outcome. This is critical for being able to implement the standards in a wide range of facilities and study designs. A performance-based approach requires professional input and judgment to achieve the specified goals at a specific facility and for a given study design.

The "Guide" is deliberately written in general terms so that its recommendations can be applied in diverse laboratories and settings; therefore, users, Institutional Animal Care and Use Committees (IACUCs), and veterinarians must use professional judgment in making specific decisions regarding animal care and use. Due to the general advice provided by the "Guide", the IACUC has a key role in the interpretation, oversight, and evaluation of the institutional animal care and use program to ensure that it complies with the general performance-based goals of the "Guide". The "Guide" uses the words "must" and "should" when describing various practices. In general, "must" is used for aspects that are imperative and mandatory; however, in some circumstances, an alternative strategy for achieving the specific goal may be acceptable but must be approved by the IACUC. In general, aspects that are described as "musts" are required for AAALAC accreditation. When auditing a laboratory for animal welfare compliance, it is important to keep in mind that while several aspects must be present, many others are left to the individual laboratory to determine how they meet the performance-based goals specified in the "Guide".

The "Guide" defines the animal care and use program broadly and includes all activities conducted at a facility that have a direct impact on the well-being of animals, including animal veterinary care, policies and procedures, personnel and program management and oversight, occupational health and safety, IACUC functions, and animal facility design and management. The responsibility for the program includes the Institutional Official, IACUC, Attending Veterinarian (AV), and animal users; however, the Institutional Official bears the ultimate responsibility for the program. The AV is responsible for the health and well-being of all laboratory animals at the facility and must be provided with sufficient authority to manage the veterinary care program. The IACUC is responsible for the assessment and oversight of the program components and facilities.

According to the "Guide", an IACUC must contain at least five members. One must be a veterinarian, one a practicing scientist, one whose primary concerns are in a nonscientific area (this person can be internal or external to the institution), and one public member who is not affiliated with the institution (e.g. lawyer, clergy, businessman, etc.). The responsibilities of the IACUC are similar to those under both the AWA and PHS Policy.

The following are some of the key functions of the IACUC:

- Reviews the animal care and use program at least every six months.
- Inspects the facilities at least once every six months.
- Prepares reports of the above IACUC evaluations and submits them to the Institutional Official.
- Makes recommendations to the Institutional Official regarding the animal care and use program, facilities, and personnel training.
- Is authorized to suspend any activity involving animals.
- Reviews and approves proposed activities and significant changes in ongoing animal-related activities. This includes approval of the study protocol and amendments. During the review, the following must be taken into consideration:
  - Procedures will avoid or minimize discomfort, distress, and pain
  - Procedures that cause more than momentary or slight pain or distress will be performed with sedation, analgesia, or anesthesia unless justified for scientific reasons in writing
  - Animals that experience severe or chronic pain or distress that cannot be relieved will be painlessly killed at the end of or during the procedure
  - Living conditions are appropriate for the species and medical care is available
  - Personnel conducting procedures are qualified and trained on those procedures
  - Methods of euthanasia are consistent with the *AVMA Panel on Euthanasia* unless justified for scientific reasons in writing
- Conducts continuing reviews of each approved, ongoing activity at appropriate intervals, as determined by the IACUC, including a complete *de novo* review at least once every three years.

All personnel involved with the care and use of animals must be adequately educated, trained, and/or qualified in the procedures they perform. This includes, but is not limited to, animal care and technical personnel, veterinarians and staff, researchers, and IACUC members.

A unique provision of the "Guide" is that an occupational safety and health program (OSHP) must be established. The OSHP helps ensure worker safety from a variety of hazards inherent with running animal studies (e.g. chemical and biological agents, physical hazards [e.g. needles], animal bites, allergens, etc.). The OSHP also helps ensure animal safety by minimizing animal exposure to potential hazards (e.g. disease transfer). The OSHP involves identifying the relevant hazards and assessing the potential risk. Managing risk involves appropriate design and operation of facilities and use of appropriate safety equipment, the development of processes and standard operating procedures to avoid or minimize the risk, and provision of personal protective equipment for employees. The OSHP must also comply with applicable laws at the federal, state, and local levels (e.g. US Occupational Safety and Health Administration [OSHA] regulations).

A few items covered by the "Guide" deserve special mention since they can impact the design of some studies:

- Emphasis is placed on choosing humane endpoints, which is the point at which pain or distress is prevented, terminated, or relieved. This can differ from the experimental endpoint, which occurs when the scientific aims and objectives have been reached. It is ideal to use humane endpoints; however, this can be difficult, especially for toxicology studies.
- The investigation of novel experimental variables is fundamental to scientific inquiry; however, this can pose challenges from an animal welfare perspective. In cases where novel variables are expected, more frequent monitoring of the animals is required to ensure their welfare.
- Physical restraint should be minimized and procedures involving prolonged restraint should be avoided unless scientifically justified and approved by the IACUC.
- The conduct of multiple major surgeries on a single animal is discouraged.
- Food and fluid intake restrictions used in some study designs should be minimized and scientifically justified.
- Pharmaceutical-grade chemicals and other substances should be used for animal procedures whenever possible.
- Processes must be in place for the continuing oversight of animal activities (post-approval monitoring). The actual procedures are left up to each facility but should ensure that animal care and use is being conducted in a manner that is consistent with the approved procedures and applicable animal welfare regulations.

The "Guide" provides relatively general information on animal facility design, housing, and management that is applicable across species since these are essential to animal well-being, quality research, and the health and safety of personnel. The following are some key considerations:

- Animal environment
  - Maintenance of body temperature within normal variations is necessary for animal well-being. Therefore, animals should be housed within temperature and humidity ranges appropriate for the species

- Ventilation must be controlled since it is important for providing adequate oxygen supply and adjusting the moisture content; removing thermal loads caused by animals, personnel, and equipment; and diluting airborne contaminants including allergens and pathogens. In addition, ventilation provides positively or negatively pressurized rooms, when needed, to ensure animal and worker safety and study integrity. In general, 10 to 15 complete room air changes per hour provide adequate ventilation of the room and most types of animal caging housed within the room. The location of air vents and diffusers should ensure even airflow throughout the room and the caging. Since the caging microenvironment can vary significantly from the room macroenvironment (e.g. airflow and removal of airborne contaminants is typically lower inside cages that only have air exposure through the tops), special attention to the impact of the caging design on the local animal ventilation must be taken into consideration (e.g. more frequent cage or bedding changes may be needed to minimize odors)
- Lighting duration, intensity, and wavelength should be considered since lighting can affect the physiology, morphology, and behavior of animals. In addition, light intensity can vary significantly in various parts of the room (e.g. bottom of a rat cage rack to the top) and means to control for the variable light intensity may need to be considered (e.g. rotating cages on the rack)
- Noise control should be considered in the facility design and placement of various species in order to minimize the impact of excessive noise on the animals and workers. In general, animal housing areas should be separated from human work locations and species that make a lot of noise (e.g. dogs, swine, nonhuman primates) should be housed in areas distant from quieter species (e.g. rats, mice, rabbits).

- Animal housing
  - The primary enclosure (e.g. cage) should provide a secure environment and be constructed of material that is nontoxic, durable, and easily cleanable. The design should prevent entrapment of animals and injury of animals and workers. Although wire mesh flooring is not prohibited, if it is used for rabbits and rodents, it is recommended that a solid resting area be provided since this can help prevent foot lesions
  - It is recommended that social animals be housed in stable pairs or groups of compatible individuals unless justified otherwise for scientific or compatibility reasons
  - Animals should be provided with adequate bedding substrate and/or structures for resting and sleeping. Contact bedding is recommended for some species (e.g. rodents) since this enhances the opportunity for species-typical behavior
  - Environmental enrichment should be used whenever it does not interfere with the scientific purposes of the study. Enrichment helps animals deal with environmental stressors leading to healthier animals. Some typical

enrichment approaches are visual barriers for nonhuman primates; nesting material, housing structures, and chewing devices for rodents; elevated shelves for cats and rabbits; toys and manipulanda for nonhuman primates, dogs, and swine. The enrichment program should be reviewed by the IACUC, researchers, and veterinarian on a regular basis to ensure it is meeting the needs of the animals and does not interfere with the scientific research

- The "Guide" provides primary enclosure sizing recommendations for species commonly used in laboratory research. However, other factors may play a role in requiring a different sized enclosure such as the age and sex, single housing versus group housing, duration of housing, and intended use. At a minimum, animals must have enough space to express their natural postures without touching the walls or ceiling, be able to turn around, have ready access to feed and water, and have sufficient space to rest away from waste
- Dogs and cats, especially when housed individually or in smaller enclosures, should be allowed to exercise and provided with positive human interaction
- Nonhuman primates should be group housed as the default and, if single housing is required, it should be scientifically justified and for the shortest duration possible. Strategies should be developed and implemented for the environmental and psychological enrichment of nonhuman primates.

- Animal care
  - Whenever feasible, methods for getting animals used to routine husbandry and experimental procedures should be used since they can reduce the stress associated with novel procedures or people
  - Wholesome, uncontaminated feed should be available on a daily basis unless required otherwise by the scientific needs of the study. Feed should be stored in clean and enclosed rooms to prevent entry of pests and the feed should be stored off of the floor to facilitate sanitation. Feeders in primary enclosures should be placed to allow easy access and prevent contamination from urine and feces
  - Potable water should be available according to each species' requirements and monitoring of water quality may be needed to ensure the quality of the water. Watering devices should be checked frequently to ensure cleanliness and appropriate operation. Automated watering systems should be flushed and/or disinfected regularly
  - Bedding materials should keep the animals dry between cage changes. It is important that bedding materials do not contain contaminants that might interfere with the health of the animals or the scientific purposes of the study (e.g. pinewood chip bedding can induce liver metabolism and alter the response to the test article)
  - The room and primary enclosures should be cleaned and sanitized at intervals that are appropriate for the species and maintain a healthy living

environment. Bedding should be changed as frequently as needed to keep the animals clean and dry and keep pollutants, such as ammonia, at low levels. In general, enclosures should be sanitized at least once every one to two weeks. Ideally, a method should be used to verify the success of sanitization

- Waste should be removed and disposed of regularly and safely
- A regularly scheduled and documented pest control program should be implemented to prevent, control, or eliminate infestations. Whenever possible, nontoxic means of pest control should be used since pesticides may interfere with the animals and studies
- Animals should be cared for by qualified personnel every day, including the availability of emergency veterinary care
- Appropriate means of animal identification should be used. Toe clipping of rodents is not recommended unless no other acceptable means are feasible
- Animal records may be helpful for ensuring and tracking animal welfare such as medical records for dogs, cats, and nonhuman primates.

The "Guide" states that a veterinary program that offers a high quality of care must be provided regardless of the number of animals or species that are maintained. The AV is ultimately responsible for the veterinary program and must be certified or have training or experience in laboratory animal science and medicine. Some aspects of the program can be carried out by other personnel so long as direct and frequent communication with the AV occurs. Key roles of the AV are providing guidance to investigators and all personnel involved with the care and use of animals about appropriate husbandry, handling, medical treatment, immobilization, sedation, analgesia, anesthesia, euthanasia, and surgery and postoperative care. The veterinary care program typically covers the following items:

- Animal procurement and transportation. Animals must be obtained lawfully and transported in compliance with the regulations.
- Preventive medicine to limit diseases and maintain healthy animals. This includes measures to identify, contain, prevent, and eradicate infections that may cause clinical disease in animals; quarantine of newly arrived animals for a sufficient period to ensure they do not spread disease to other animals or personnel; separation of species and strains to prevent disease transmission and stress; and at least daily observations of the animals to ensure they remain healthy.
- Clinical care and management to ensure animals are well cared for, remain healthy, and receive proper emergency care. This also includes ensuring that animals experiencing adverse effects in studies are appropriately cared for.
- Surgery and postoperative care.
- Pain and distress recognition and prevention or alleviation whenever possible.
- Appropriate use of anesthesia and analgesia.

- Euthanasia in accordance with the AVMA Guidelines on Euthanasia unless justified otherwise for scientific or medical reasons. Euthanasia must be performed by personnel who are skilled in the methods for the given species and in a compassionate manner.

The "Guide" provides general advice on the construction of the facilities since a well-planned, constructed, maintained, and managed facility facilitates efficient, economical, and safe operations, which in turn enhances animal welfare. Animal housing areas should be separated from personnel areas such as offices, break rooms, and conference rooms. This can be accomplished by various means such as physically separate buildings or isolated areas of a single building using appropriate corridors, barriers, and entry locks. Areas housing animals should be dedicated for animal housing and should not be multipurpose laboratories where other scientific functions are routinely conducted. A facility may use a centralized design where support, care, and use areas (e.g. cage washing, feed and caging storage) are adjacent to the animal rooms whereas others may use a decentralized design or a combination of the two. A decentralized design poses additional challenges since animals and equipment must be transported between the animal rooms and the use areas increasing the risk for transport stress and exposure to disease.

Most animal facilities will need space for:

- Animal receipt and quarantine
- Animal housing, care, and sanitation
- Separation of species and/or isolation of projects
- Storage areas for equipment, feed, and bedding
- Specialized laboratory space such as necropsy or general purpose procedure rooms
- Equipment for washing cages and associated supplies
- Storing waste and disposal of carcasses.

Ideally, the entire premises of the animal facility should have restricted vehicle and personnel access to the property followed by restricted personnel access to buildings. Animal housing facilities should be separately secured and have limited access to only authorized personnel, such as through electronic card-key access. Appropriate security monitoring procedures should be used for the entire facility such as alarms and security cameras.

The following are some considerations for general construction of the animal areas:

- Corridors should be wide enough to allow efficient movement of personnel and equipment and floor–wall junctions should be designed to facilitate cleaning
- Double door entries should be considered for high noise areas such as cage wash and dog, swine, or nonhuman primate areas
- Doors to animal rooms should be large enough to allow efficient passage of caging and equipment and the doors should fit tightly in their casings to prevent vermin entry

- Exterior windows are not recommended in animal areas since they pose a security risk, can potentially alter the photoperiod, and can make temperature regulation more difficult
- Floors should be constructed to withstand moisture, be impact resistant, and have none to a minimal number of joints to facilitate cleaning
- Floor drains can be used to help facilitate cleaning of animal rooms and removal of waste, especially with larger animals such as dogs, swine, and non-human primates
- Walls and ceilings should be constructed to withstand moisture and be easily cleaned and sanitized. Suspended ceilings are not recommended
- The heating, ventilation, and air conditioning (HVAC) system should be able to maintain the desired temperature within $\pm 1°C$ and humidity within 30–70% and provide the necessary number of air changes per hour. The animal rooms should be monitored to ensure the environment is appropriately maintained
- A lighting control system should be used to provide consistent diurnal lighting
- A back-up power system should be available to maintain critical services such as the HVAC during power failure. The back-up system should be tested regularly to ensure it is functioning appropriately
- Specialized areas may be needed for surgery, barrier facilities for the use of or to exclude infectious agents, imaging facilities, among others.

REGULATIONS ESTABLISHED BY THE USDA UNDER THE AWA

As mentioned previously, the AWA specifically excludes mice of the genus *Mus* and rats of the genus *Rattus* bred for research. Therefore, a facility conducting nonclinical studies utilizing species under these two genera exclusively may not be subject to the AWA. If the facility is subject to the AWA, it must register with the USDA every three years. For these facilities, the USDA ensures compliance with the animal welfare regulations through USDA-driven inspections and self-monitoring. Facility self-monitoring is driven by three key components: the Institutional Official, typically the head of the facility, the AV, and the IACUC. Together, these three entities ensure that the animal care and use program at the research facility provides for the humane care and use of animals and compliance with the regulations. The Institutional Official is ultimately responsible for the proper functioning of the animal care and use program; however, the day-to-day functioning of the animal care and use program is primarily done by the IACUC and the AV and his or her staff.

An annual report must be prepared and submitted to the USDA each year. An important component of the report is a listing of the number of animals used in the following pain categories:

- Animals being bred, conditioned, or held but not yet used. These are also referred to as *USDA category B* animals since they are reported in column B of the USDA annual report form
- Animals upon which teaching, research, experiments, or tests were conducted involving no pain, distress, or use of pain-relieving drugs. *USDA category C* animals

- Animals upon which experiments, teaching, research, surgery, or tests were conducted involving pain or distress and for which anesthetic, analgesic, or tranquilizing drugs were used. *USDA category D* animals
- Animals upon which experiments, teaching, research, surgery, or tests were conducted involving pain or distress and for which the use of anesthetic, analgesic, or tranquilizing drugs would have interfered with the scientific purpose of the study. An explanation of the procedures producing pain or distress and reasons why drugs were withheld must be attached to the annual report. *USDA category E* animals.

A great deal of emphasis is placed on animals within categories D and especially E since these animals are most likely to experience significant pain or distress. Because there are increased justification and reporting requirements for category E animals, the laboratory may ask the Contracting Scientist to provide sufficient background information and scientific justification.

The USDA conducts unannounced annual inspections. During an inspection, the USDA official can examine and copy records that are required to be maintained by the regulations (e.g. IACUC, animal procurement, care records, etc.), inspect the facilities, property, and animals, and document by taking photographs or other means conditions and areas of noncompliance. The laboratory should notify the Contracting Scientist whenever an inspection has taken place and if there were any concerns raised during the inspection, especially those that directly impact the Contracting Scientist's studies.

POTENTIAL CONFLICTS BETWEEN THE ANIMAL WELFARE REQUIREMENTS, GLPs, AND OTHER STUDY REQUIREMENTS

As mentioned previously, some regulatory requirements may conflict with each other or make it difficult to determine who has the ultimate say in certain study-related decisions. It is important to be aware of these since they can lead to potential conflict between the Contracting Scientist, Study Director, AV, IACUC, and others. These issues often come up when unanticipated adverse effects occur in a study and a quick course of action needs to be taken to ensure animal welfare. Although these effects cannot be predicted, an appropriate course of action with a clear outline of responsibilities can be established before the start of the study so that all parties understand their roles when an unanticipated effect occurs.

Under the GLPs, the Study Director has overall responsibility for the technical conduct of the study and is the single point of control. This implies that the Study Director needs to make all decisions regarding the conduct of the study. However, under the AWA and the "Guide", the AV must have the authority to make decisions about animal care and use. Since decisions about animal care and use are entwined with study conduct, there is a conflict about who has the authority and responsibility to make certain decisions involving animal care and use, especially when they impact the scientific conduct of the study. Ideally, the AV, Study Director, and Contracting Scientist should work together to arrive at appropriate solutions. However, it should also be clearly established before the study starts and within the protocol, who has the ultimate authority if a consensus cannot be reached. For example, an animal may be experiencing severe pain and distress that was unexpected. The Study Director and Contracting Scientist want to keep the animal on study but the AV determines that the animal is close to being moribund and should be euthanized. The majority of laboratories will default to the decision of the AV to avoid any animal welfare issues; however, the Study Director is ultimately responsible for the scientific conduct of the study, presenting a potential conflict.

Humane endpoints should be included in the protocol so that clear criteria for euthanasia, supportive care, or removal from the study are established. The protocol should include criteria for initiating euthanasia such as the degree of physical or behavioral deficit or tumor size that will enable a prompt decision to be made by the veterinarian and the investigator to ensure that the end point is humane and the objective of the protocol is achieved. This can be challenging for some studies, such as toxicology studies, since the potential induction of serious adverse effects may be a scientific requirement of the study. Also, some studies may require a certain number of animals to survive until a certain age and/or the planned necropsy in order to have a valid study. If animals are euthanized prior to this, the study may have to be repeated which poses a dilemma between keeping sick animals on the current study until the designated timepoint is reached versus using additional animals and repeating the study.

The use of analgesics for animals in pain or distress is often a contentious point for studies. The AV and IACUC are required to prevent or eliminate pain or distress but analgesics can interfere with the response of the animals to the test article making the study scientifically invalid. Depending on the laboratory, IACUC and AV, the amount of written scientific justification for withholding analgesics can vary greatly and this should be taken into consideration. This also applies to environmental enrichment since the "Guide" recommends enrichment for all animals but the scientific needs of the study may require no enrichment since the enrichment is another study variable that could impact the interpretation of the study findings. If no enrichment is used, it must be scientifically justified and approved by the IACUC.

ANIMAL WELFARE CHECKLIST

The following checklist can be used to help ensure that a facility is meeting the necessary mandated and voluntary animal welfare requirements. Since the GLPs cover some animal welfare issues and animal welfare is intimately tied to the study design and day-to-day running of the study, some of the items are also covered in the checklists provided in the chapters on GLPs and study design.

ANIMAL WELFARE	YES	NO	N/A
Certifications			
1. Laboratory is AAALAC accredited			
When was the last inspection?			
Were there any findings or recommendations (review report)?			
Has the laboratory ever lost AAALAC accreditation?			
2. How many veterinarians are board-certified in laboratory animal medicine?			
ACLAM			
None			
3. How many technicians and supervisors are AALAS certified?			
CMAR			
ALAT			
LAT			
LATG			
None			
General Regulatory Compliance			
1. In the USA, does the laboratory use USDA-regulated species and is the laboratory registered with the USDA?			
2. Does the laboratory conduct US Department of Health and Human Services funded research?			
Do they comply with the PHS Policy and do they have an approved Assurance (review a copy)?			
3. Does the laboratory run studies according to the "Guide" unless there is sufficient scientific justification?			
4. Facility has:			
IACUC			
Institutional Official			
Attending Veterinarian			
IACUC			
5. IACUC membership meets the "Guide" requirements:			
At least 5 members			
One a veterinarian			
One a practicing scientist			
One whose primary concerns are not scientific (e.g. lawyer, clergy)			
One not affiliated with the laboratory			
6. How often does the IACUC meet?			
7. Reviews at least every 6 months the laboratory's animal care and use program (obtain and review a copy of the last review)			
Any major issues/concerns?			
8. Reviews and inspects at least every 6 months the animal facilities (obtain and review a copy of the last review)			
Any major issues/concerns?			
9. Prepares reports of the reviews/inspections and submits them to the Institutional Official			
Report contains significant and minor deficiencies			
Report contains a schedule and plan for correcting each deficiency			

ANIMAL WELFARE	YES	NO	N/A
10. Reviews and approves study protocols and amendments from an animal care and use perspective			
11. What is the protocol/amendment review process (how many people review each protocol, how is approval given)?			
12. Makes recommendations to the Institutional Official about the animal care program, facilities, and personnel training			
13. Has authority to suspend activities involving animals For suspended activities, they are reviewed by the IACUC and Institutional Official and appropriate corrective action is taken			
14. A process is in place for continuing reviews of approved, ongoing activities to ensure animal care and use procedures are being conducted in accordance with the approved protocol (e.g. phase inspections by IACUC, veterinary staff, and/or compliance liaison)			
Protocol			
1. Protocol contains the following information: Identification of the species and number of animals to be used Description of the use of the animals Scientific justification for procedures that may cause discomfort or pain Description of euthanasia methods			
2. Protocol contains humane endpoints for ending the study, whenever possible. These are endpoints for which pain or distress is prevented, terminated, or relieved			
3. Procedures will avoid or minimize discomfort, distress, and pain unless scientifically justified What is the scientific justification, if applicable?			
4. Procedures that may cause more than momentary slight pain or distress will be performed with sedatives, analgesics, or anesthetics unless scientifically justified in writing Attending Veterinarian has been consulted on these procedures What is the scientific justification, if applicable?			
5. Alternatives have been considered to procedures that may cause more than momentary slight pain or distress Written justification has been provided along with method/sources for determination (e.g. literature databases searched)			
6. Study does not unnecessarily duplicate previous experiments			
7. Animals experiencing severe or chronic pain or distress are euthanized at the end of or during the procedure			
8. Studies involving novel experimental variables have increased frequency of animal monitoring to ensure animal welfare			
9. Physical restraint is minimized and prolonged restraint is scientifically justified and approved by the IACUC What is the scientific justification, if applicable?			

ANIMAL WELFARE	YES	NO	N/A
10. Feed and fluid restrictions are scientifically justified What is the scientific justification, if applicable?			
11. Pharmaceutical-grade chemicals and other substances are used for animal procedures (e.g. anesthesia), whenever possible			
12. Surgery is conducted aseptically			
13. Surgery for non-rodents is conducted in dedicated facilities			
14. Euthanasia is humane and conducted in accordance with the AVMA Guidelines on Euthanasia unless justified for scientific or medical reasons Technicians are trained and experienced in method How is death assured? What is the scientific/medical justification, if applicable?			
Personnel and Training			
1. Personnel conducting procedures and animal care are qualified and trained in those procedures			
2. Personnel have been trained in: Humane methods of animal maintenance and experimentation Proper use of anesthetics, analgesics, and tranquilizers How to investigate alternatives that could replace, reduce, or refine proposed research			
3. Personnel know how to report deficiencies in animal care and treatment There are no repercussions for reporting			
4. There are enough employees to carry out a level of animal husbandry that ensures animal welfare			
5. Staff conducting animal husbandry and care are supervised by an individual who has the knowledge, background, and experience in proper husbandry and care to supervise others			
Attending Veterinarian			
1. Is certified or has training or experience in laboratory animal science and medicine			
2. Has authority to ensure the provision of adequate veterinary care and oversee other aspects of animal care and use			
3. Is a voting member of the IACUC or the role has been delegated to another veterinarian			
4. Establishes veterinary care program that: Provides appropriate facilities, personnel, equipment and services Uses appropriate methods to prevent, control, diagnose, and treat injuries and disease Conducts daily observation of all animals to assess their health and well-being Provides guidance to principal investigators (Study Directors and Sponsors) on animal care and use Provides guidance on appropriate use of anesthesia and analgesia Ensures adequate pre- and post-procedural care			
5. Maintains large animal medical records and records are updated and complete			

ANIMAL WELFARE	YES	NO	N/A
Occupational Safety and Health Program (OSHP)			
1. An OSHP is in place			
2. Personnel protective equipment (PPE) is used. Describe PPE for: Animal rooms Test article handling General laboratory			
3. Are there any unique hazards (e.g. biohazard, highly toxic test article)? Have they been appropriately identified by the facility? Are procedures in place to minimize potential risk?			
4. Are there special procedures needed for the study (e.g. use of shower-in/shower-out, foot baths, clean/dirty corridors)? Are personnel aware of and trained in the procedures?			
Animal Care and Housing			
1. The living conditions are appropriate for the species and medical care is available Temperature and humidity is controlled and appropriate for species Ventilation of room and primary enclosures is adequate Lighting duration, intensity, and wavelength is appropriate for species Noisy species (e.g. dogs and swine) are distant from quieter species (e.g. mice and rats)			
3. Primary enclosures meet the minimum size requirements specified in the "Guide"			
4. Floors of the primary enclosure protect the animal's feet from injury If wire bottom cages are used, are areas provided where the animals can get off of the flooring (e.g. solid resting area)?			
5. Design of the primary enclosure prevents animal injury (e.g. no sharp edges or entrapment risks)			
6. Adequate bedding and/or structures for resting and sleeping are provided Bedding materials do not contain contaminants that interfere with the study			
7. What is the frequency of: Bedding changes Primary enclosure cleaning Primary enclosure sanitation Room cleaning Room sanitation			
8. How are enclosures cleaned and sanitized (e.g. cage wash)?			
9. What chemicals are used for cleaning and sanitation?			
10. How is sanitation verified (e.g. cultures) and how often? Primary enclosures Animal rooms			
11. Cat enclosures have an elevated resting surface and a litter box			

ANIMAL WELFARE	YES	NO	N/A
12. Animals are group housed as long as they are compatible and it does not interfere with the scientific purpose of the study			
If single housing, written justification has been provided			
13. Environmental enrichment is used and it is beneficial			
Plan is in place for all species			
Written scientific justification is provided for exceptions			
14. Plan is in place for providing dogs with the opportunity for exercise			
If single housing dogs without group exercise, is the primary enclosure large enough to meet the exercise requirement?			
15. Plan is in place for the environmental enhancement of nonhuman primates in order to promote their psychological well-being. The plan addresses:			
Social grouping			
Environmental enrichment			
Special attention for some primates (e.g. infants)			
The use of restraint devices			
16. Animals are habituated to study procedures (e.g. sham dosing or handling during acclimation)			
17. Feed is available on a daily basis			
Bulk feed is stored in clean, enclosed rooms and off of the floor			
Feed storage in animal rooms is in sealed containers and appropriately labeled (e.g. with type and expiration date)			
Animal feeders allow easy access and prevent contamination from urine and feces			
18. Potable water is readily available			
Water source (e.g. well, municipal)			
Water is monitored for quality			
Watering devices are checked frequently for cleanliness and appropriate operation			
Automated watering systems are flushed and/or disinfected regularly			
19. Animals are cared for every day			
20. Emergency veterinary care is available			
21. Animals are appropriately identified (e.g. tattoo, RFID chip)			
Toe clips and ear punches/clips are not used unless no other feasible means			
22. Euthanasia is conducted by personnel experienced in the procedure			
Facilities			
1. There is space for:			
Animal receipt and quarantine			
Animal housing, care, and sanitation			
Separation of species and/or isolation of projects			
Storage areas for equipment, feed, and bedding			
Specialized laboratory space (e.g. necropsy or general purpose)			
Equipment for washing cages and associated supplies			
Storing waste and disposal of carcasses			

ANIMAL WELFARE	YES	NO	N/A
2. Construction of floors, walls, and ceilings prevents damage from animals and routine use and can be easily cleaned and sanitized			
3. Animal rooms lack exterior windows or skylights If door has window, is the corridor light cycle similar to the room?			
4. Facility has restricted access to: Entire site (e.g. vehicles and people) Individual buildings Animal housing areas			
5. Facility has security system, is actively monitored, or security personnel			
6. What are guest policies (e.g. sign in/out, escort) and PPE requirements?			
7. Back-up power is available and can run animal area HVAC system Does it come on automatically? How often is it tested?			
8. Pest control program is in place Are only nontoxic means used? If pesticides are used, what are they and do they interfere with the study?			
Records and Reporting			
1. Retains the following records: Minutes of IACUC meetings Records of proposed animal activities (i.e. study protocols and amendments) and whether IACUC approval was given or withheld Records of semiannual IACUC reports and recommendations			
2. For USDA-regulated facilities, files annual report (obtain a copy and review the most recent report) Annual report contains sufficient written justification for procedures producing pain or distress and why drugs were withheld (Category E animals)			
3. For PHS OLAW-regulated facilities, files annual report (obtain a copy and review the most recent report). The report contains: Any change in AAALAC accreditation (category change) Changes in the animal care and use programs Change in IACUC membership Dates that IACUC conducted its semiannual evaluations and submitted them to the Institutional Official			
4. For PHS OLAW-regulated facilities, OLAW is promptly notified of the circumstances and actions taken for: Any serious or continuing noncompliance with the PHS policy Any serious deviation from the Guide Any suspension of activity by the IACUC Obtain a copy and review any such notifications			

ANIMAL WELFARE	YES	NO	N/A
Inspections			
1. For USDA-regulated facilities, obtain and review a copy of the most recent USDA inspection and the laboratory's responses			
2. For PHS OLAW-regulated facilities, obtain and review a copy of the most recent annual report, any reports of suspended activities, and any communications between OLAW and the laboratory			

Chapter 5

Laboratory Selection

Karen VanLare BA, RVT* and William F. Salminen PhD, DABT, PMP†
*Novartis Animal Health, Greensboro, NC, †PAREXEL International, Sarasota, FL

Key Points

- Screening and qualifying a laboratory is the first step for obtaining a successful study
- Laboratories often have different strengths and weaknesses that can impact the conduct or reporting of the study
- It is important to determine a laboratory's experience with your given study design

This chapter focuses on selecting a new laboratory to run nonclinical studies. Methods for screening laboratories will be presented along with ways to assess their past performance. At the end of the chapter, a monitoring checklist will be provided which will help the reader screen new laboratories when it is used in conjunction with the checklists provided in the previous chapters. Laboratory selection is critical to the success of any study since the laboratory will be conducting most, if not all, of the various study functions leading to a final report that can be used to support a regulatory submission. Therefore, it is critical that the laboratory has the experience in conducting the given type of nonclinical study, they follow the protocol and adhere to regulatory requirements (e.g. Good Laboratory Practices [GLPs]), and they generate sound data that are accurately summarized in the final report. The last thing a Contracting Scientist wants is a study that was poorly conducted and/or the final report does not accurately reflect the raw data since this could jeopardize the overall regulatory submission. As mentioned in the introductory chapter, some laboratories excel at certain aspects of study conduct but may fall short on others. It is important to assess thoroughly the laboratory's capabilities and strengths and weaknesses so that you fully understand how the study will be run and whether or not the laboratory can realistically meet your expectations.

Nonclinical Study Contracting and Monitoring. http://dx.doi.org/10.1016/B978-0-12-397829-5.00005-3

As with companies, each laboratory has its own personality. Some laboratories are very compartmentalized and run with a very rigid and formal structure. Others have a more fluid feel with groups working closely together and personnel even conducting multiple diverse study functions to complete the study. There is no "right" structure and it is a personal preference as to the type of laboratory a Contracting Scientist likes to work with. Often, a Contracting Scientist will conduct studies at multiple laboratories and the laboratories will range in their structure and working style. What is important is for the Contracting Scientist to understand the laboratory's working style so that they clearly understand how a study is run within the facility and how personnel from diverse functions work with each other (e.g. how well do the in-life and necropsy groups coordinate activities, do the Quality Assurance Unit [QAU] and Study Director work in a cooperative manner). This can help the Contracting Scientist understand what study aspects they may need to monitor closely to ensure that all the study functions are conducted and reported correctly.

CONTACTING AND PRELIMINARY SCREENING OF A NEW LABORATORY

There are many nonclinical laboratories to select from and it can be a daunting task trying to determine which laboratory has the necessary capability and expertise to conduct your study. Laboratories range in size from ones that conduct one or two studies at a time to ones that run hundreds of studies at multiple locations throughout the world. The first step in screening laboratories is determining if they have the basic capability to conduct your study. There are a variety of methods you can use to determine a laboratory's basic capabilities. Websites often provide general information but it may be too generic to determine if they can meet your requirements since companies often try to keep a low profile to limit animal rights activist inquiries and protests. Therefore, you often need to contact the laboratory and meet with a sales and/or technical representative from the laboratory. A great place to meet a lot of laboratories and obtain detailed capability information is at scientific meetings such as the annual Society of Toxicology and American College of Toxicology meetings. Most nonclinical laboratories will have booths at the major scientific meetings and you can obtain detailed literature from the laboratories that is a lot more specific than the information on their website and you can meet with both sales and technical representatives. If you cannot meet a laboratory at a scientific meeting, then having them visit you is another great way to get to know a laboratory. Most laboratories are more than eager to try and win your business and an integral part of this is sending a representative to visit you and go over their capabilities. You should also screen laboratories based on peer recommendations and experience. Colleagues may have used a laboratory you are considering and their first-hand experience is invaluable in determining not only a laboratory's basic capabilities but also their working style and quality.

Meeting with a Sales and/or Technical Representative

Laboratories are in the business of making money and they do this by keeping the laboratory as full to capacity as possible. The majority of laboratories are continually seeking new clients to keep their laboratories full. It is important to keep this in mind since a laboratory may oversell themselves, either on capabilities or price, just to get their foot in the door. Be wary of laboratories that say they can run any type of study you need. Laboratories will often specialize in certain types of studies and, in general, you should run your study at a laboratory that has extensive experience running that type of study. Some studies may be so unique that no laboratory has experience with that type of study; however, in those circumstances, you can select a laboratory that has the closest experience and is also clearly capable of learning and implementing any novel procedures.

When you finally decide to meet with the laboratory, it is best to meet with both sales and technical representatives. If the sales person came up through the technical ranks, he may have the necessary background for discussing the technical requirements. The sales person will provide you with the big picture overview of the laboratory and is typically your main contact for study pricing and ensuring that your future inquiries are directed to the right people and answered in a timely manner. The technical person can provide quick answers to many of your study-specific questions and provide insight into how the laboratory generally runs their studies. By meeting with both people at the same time, you can get a good feel for the laboratory, their working style, and whether they can meet your basic requirements. During your meeting, you should get a general overview of the laboratory's capabilities along with the types of studies they typically run or specialize in. They will often provide you with standard pricing for their more routine studies, which is valuable for future pricing comparisons between laboratories. However, since competition is strong between laboratories, you should always consider the pricing as a starting point and negotiate from there since you can often get price breaks if the laboratory is not at capacity and is looking to fill rooms or if you are placing a battery of studies with the laboratory.

After the general overview, you should go over your general needs and study requirements. It is very helpful if you can go over the specific studies you need conducted. The laboratory representatives will be able to tell you if they can conduct the studies and you can probe them about the specific details. Having the technical representative present will allow you thoroughly to probe them about specific study details and how the laboratory handles certain study items so you can determine if they truly have experience with your study designs or if they are overstating their abilities. Many sales representatives will take you out to lunch or dinner as part of their visit. It is best to reserve this until after the hard discussions so that you can ask detailed questions without distraction. The following are some typical items that can be covered:

- Experience
  - Does the laboratory have experience with the given species?
    - How many studies do they currently have with this species?
    - Where are the animals sourced from (e.g. external breeder, in-house colony)?
  - Does the laboratory have experience with the given study design?
    - How many studies of this design have they run?
    - When is the last time they ran a study with this design?
  - Does the laboratory have experience with specific and unique study requirements (e.g. blinding of the technicians to dose groups, sperm collection and analysis, telemetry)?
  - Does the laboratory have experience with the specific reporting agency or center (e.g. Food and Drug Administration Center for Drug Evaluation and Research [FDA CDER] versus FDA Center for Veterinary Medicine [FDA CVM], Environmental Protection Agency [EPA], US Department of Agriculture [USDA], European Medicines Agency [EMA]).
- Capacity and timing
  - What is the laboratory's capacity (e.g. number of animal rooms for various species)?
    - What is lead time needed for starting a study?
    - When does test material have to be provided?
    - When does the protocol have to be finalized?
    - How soon can animals be obtained?
- Personnel
  - How many personnel are there and what are their credentials (e.g. education level and certifications)?
    - Study Directors
    - In-life technicians
    - Formulations
    - Clinical pathology
    - Necropsy-pathologists and technicians
    - QAU
    - Report writing
  - Can the Contracting Scientist select a specific Study Director to work with (e.g. one with extensive experience in the given study design)?
  - For masked (blinded) studies, are there enough personnel to conduct the study if masking is separated by function?
- Test article and formulations
  - What are the test article handling procedures (e.g. receipt, distribution, accountability)?
    - Can the laboratory prepare the formulations, when applicable?

- Can they also conduct analysis for stability, homogeneity, and concentration?
        - Do they have the capacity and can they retain samples properly?
- Capabilities
    - What portions of the study can be conducted at the laboratory?
        - Formulations
        - Dosing and in-life functions
        - Clinical pathology
        - Necropsy
        - Other specific requirements (e.g. pharmacokinetic analysis, ECGs).
- Protocol
    - What is the process for developing a protocol (e.g. does the Study Director provide a draft after a teleconference to discuss the study)?
        - Can you use a company-specific protocol format, if needed?
    - How are protocol amendments handled and approved (e.g. does the Contracting Scientist provide written approval)?
    - Does the CRO have a pre-study protocol meeting with the study team to discuss logistics and clarify any outstanding questions prior to study start?
- Study conduct
    - How are animals identified (e.g. tattoo, radio frequency identification [RFID])?
    - How are data collected for various study functions (e.g. manual and/or computer)?
        - If computer, what system is being used and is it 21 CFR Part 11 compliant?
    - Is there a method for obtaining and reviewing data generated to date (e.g. secure website)?
        - Does the Study Director provide routine updates and what is the format (e.g. written study summaries)?
    - Does the Study Director immediately contact the Contracting Scientist about any significant study-related problems or sick/dead animals?
        - How often are less serious study deviations reported to the Contracting Scientist?
    - How do personnel communicate with the Study Director? For example, Study Director Notifications, Notes to File, eNotes, etc.
- QC (Quality Control)
    - Once the data are collected, is there a process for reviewing the data for errors (prior to QAU review)?
    - Is the raw data reviewed for errors prior to sending for interpretation (statistical analysis, Study Director review/interpretation, Sponsor Representative review, cardiologist review/interpretation, etc.)?
- QAU
    - What is the structure of the QAU (e.g. how many personnel, reporting structure)?
        - What are the typical phase inspections?

- What is the auditing process for raw data (e.g. what percentage is audited)?
- What is the auditing process for the draft and final reports (e.g. auditing against the raw data)?
- How will other Test Sites report to the Test Facility QAU?

- Reporting
  - How are the data compiled into a final report (e.g. report writing group in conjunction with the Study Director)?
    - How are contributing scientist reports and non-GLP reports handled?
    - What is the draft report review process and how are changes communicated between the Contracting Scientist and the laboratory (e.g. paper or marked-up electronic file)?
    - What is the quality of the draft report (e.g. includes all reports and data have been thoroughly quality checked)?
    - Can you use a company-specific report format, if needed?
  - What is the format of the final report (e.g. paper and/or electronic file)?
    - What is the reporting timeline and what is it based on (e.g. time from the last in-life function or time from the end of the reading of the histopathology slides)?
  - Is the draft report audited prior to Sponsor Representative review?
- Other
  - Who is the main contact for:
    - Ongoing studies (e.g. Study Director)
    - New studies (e.g. sales representative)
  - Does the laboratory have a price list for routine study designs?
  - Does the laboratory have reporting timelines for routine study designs?
  - Can study reporting be expedited for an additional fee?

After the meeting is completed, you should have enough information to determine if the laboratory has the general capabilities successfully to run your study. This is the first step of the screening process and the next step involves visiting the laboratory, meeting with key laboratory personnel that will be integral to running your study, and thoroughly auditing their procedures and processes.

VISITING AND AUDITING A NEW LABORATORY

Once you have decided to investigate a laboratory's capabilities in more detail, an excellent way to obtain a good overview of their capabilities, processes, procedures, quality, and working style is to visit the laboratory. Depending on what aspects will be audited by the Contracting Scientist, one to three days is usually sufficient to conduct a thorough assessment of the laboratory's operations and regulatory compliance. If the Contracting Scientist's company has regulatory compliance and animal welfare representatives that will audit the laboratory, the Contracting Scientist can focus on the laboratory's ability to conduct the study from more of a scientific perspective and a one day visit will likely be sufficient.

However, if the Contracting Scientist needs to conduct the GLP and animal welfare audits, up to three days may be needed to audit thoroughly the laboratory's processes and procedures. Conducting a thorough assessment before any study is placed at the laboratory increases the chances that the laboratory will meet your requirements and will also set a strong tone with the laboratory about the type of quality study you require.

Once you have scheduled a visit with the laboratory, you should ask the laboratory for copies of their most recent regulatory filings and inspections (e.g. FDA GLP, EU GLP, USDA Animal Welfare [annual inspection and annual reporting]) and responses they provided to any findings or significant deviations. Having these prior to the visit will give you time to review them and prepare any questions. If the laboratory is reluctant to release copies of the reports and responses, this is an immediate red flag that the laboratory may be trying to downplay recent findings. Also, if you are leery that the laboratory may not be providing all of the most recent reports or are omitting some reports, in the USA you can submit a freedom of information request to obtain copies of the various reports. Details on making freedom of information requests from the FDA and USDA can be found at:

- http://www.fda.gov/RegulatoryInformation/FOI/HowtoMakeaFOIARequest/default.htm
- http://www.aphis.usda.gov/foia/how_to_submit_a_foia_request.shtml

Some information, such as FDA warning letters arising from serious findings during laboratory GLP inspections, can be found immediately online. FDA warning letters are available at: http://www.fda.gov/iceci/enforcementactions/warningletters/default.htm

Some laboratories may offer to pay for your initial visit or portions of your stay (e.g. a hotel). Before accepting any such offers, you should check with your legal and ethics office to make sure that it does not go against any company policies. In addition, some laboratories may provide complimentary gifts (e.g. gift baskets or company apparel and items) when you visit and take you out to lunch or dinner. Although most companies allow you to accept gifts and meals of modest value, it is always best to check with your legal and ethics office to ensure you are abiding by company policies.

In preparation for your visit, you should tell the laboratory exactly what you want to accomplish during your visit. For example, you should tell them that you want a tour of the facilities, which will include conducting a detailed audit, you want to meet with the heads of the key groups (e.g. Study Directors, formulations, in-life technical operations, necropsy, QAU, attending veterinarian), you want to audit their Standard Operating Procedures (SOPs) and other facility documentation, and you want to go over your study needs in detail. This will ensure that they have the relevant information available for review and that you can meet with key people. If you have one or more studies that you are contemplating placing at the laboratory, it is best to provide the laboratory with a detailed study outline (this is reviewed in the next chapter). This will help guide the discussions and allow you to assess fully the laboratory's ability to conduct the specific study. If you provide sufficient detail in the study outline, the laboratory can make sure that you meet with key people responsible for various critical or unique study functions (e.g. formulations, in-life dosing, pharmacokinetic analysis). This will also allow them to provide you with cost estimates for the study that are likely to be similar to the price estimate after the final protocol is developed. Prior to your visit, the laboratory should send you a tentative schedule so you can assess if they have scheduled meetings with all of the appropriate people and given you sufficient time to conduct your reviews. If you are going to disclose confidential information (e.g. overview of studies needed to support a new drug registration), you should establish a confidential disclosure agreement (CDA) with the laboratory before releasing any proprietary information and make sure that you clear any disclosures with your company. Details on establishing a CDA are presented in a later chapter.

The following information assumes you will be conducting an audit of the laboratory from study design/scientific, GLP, and animal welfare perspectives. Prior to your arrival at the laboratory, you should review the various checklists so that you are familiar with the various items that need to be assessed. Even if other people will conduct the GLP and animal welfare audits, it may be beneficial for the Contracting Scientist still to complete the GLP and animal welfare checklists, or at least relevant portions, since he/she may have a different viewpoint for some of the items that may raise concerns and would not necessarily be caught by the other inspectors.

On the day of your arrival, you should assess the location and security procedures for accessing the facility. Many animal facilities are in remote locations and this is generally preferable since it can limit public visibility and potential protests from animal rights activists. It is preferable if the entire site has limited access (e.g. fenced perimeter with key-carded vehicle access); however, at a minimum, the individual buildings should have restricted access to authorized personnel only. Once you arrive at the reception area, assess the procedures used for signing in, orienting (e.g. instructions on potential hazards and personnel protective equipment), and escorting visitors. In order to keep competitors from knowing that you may be placing studies at the laboratory, if this is important for confidentiality or competition reasons, make sure that a sign in/out procedure is used that prevents other guests from knowing you visited (e.g. a paper sign in/out guest book should not be used).

Most laboratories will provide a quiet dedicated room where you can settle in and conduct your reviews. You may also meet with the key study and facility personnel in that room. Once you have settled in, the hard work begins. Conducting a thorough assessment of a new laboratory entails a lot of work and attention to detail. Once completed, you will have a lot of information that will help you determine the ability of the laboratory to run your study and meet your expectations. In addition, a thorough audit will set a strong tone with the laboratory that you have high expectations for every study and that you will play an active role in running and monitoring each study.

A tour of the facility should be one of the first items on your agenda. You should bring along a copy of the floor plan, if possible, so that you can make notes about different areas. You should assess the general layout and cleanliness of the facility and security and personal protective equipment procedures, particularly for animal room access. You should also assess how study information is presented on the animal rooms and study-related documentation to ensure coded means are used so that competitors cannot identify your studies and/or test article when they walk through the facility. As you walk through the facility, you should fill out the relevant checklists (e.g. laboratory selection [at the end of this chapter], study design, GLP, and animal welfare). You should spend as much time as needed during the walk through and not be rushed.

You should make it clear to your host that you will be doing a thorough audit of their facility and procedures and this will take more time than a simple walk through. Since there are a lot of items on the checklists that are assessed separately from the walk-through, you may want to highlight the items that need to be assessed during the walk through so that you can easily identify them. You will want to visit all of the major areas of the facility including but not limited to:

- Animal receiving area
- Animal rooms
- Animal veterinary treatment area
- Cage and equipment washing and storage
- Feed and bedding storage
- General supplies storage
- General receiving, particularly for test articles
- Formulations, including test article storage and mixture preparation
- Analytical laboratory (e.g. for mixture analysis and/or pharmacokinetic assessments)
- Clinical pathology laboratory
- Necropsy
- QAU and archives
- Other specialized areas.

When you are done with your tour, you should have a good base for assessing the layout and functioning of the facility. If there are outstanding questions or areas that were missed, do not hesitate to ask to go back to the relevant areas so that you can complete your audit.

At some point during your visit, you should present your general study requirements and go over one or more study outlines. This gives the laboratory a good idea of what you are looking for and if they can deliver what you need. It is best if you can provide the detailed study outline to the laboratory prior to your visit so that they are familiar with the study and have relevant questions prepared. There are often many questions that are raised during the review of the studies and this is a great time to get a feel for the working relationship between you and the laboratory personnel. Since many people are involved with the conduct of the study, all of the key personnel should be present during the study review so that the right questions can be raised and answered from both sides. You should go over the details of the study to ensure that everyone understands the study design and any special requirements (e.g. the protocol may need to be submitted for regulatory approval before the study starts). You can also work out timing logistics for the study such as when does test material have to be supplied, when and where are animals purchased, and when will the formulations and analytical work be completed relative to study start. Although less important from the viewpoint of running a study, you can also give an overview of the project the studies are being conducted to support (e.g. drug ABC is being developed to treat Parkinson's disease). This helps get the laboratory interested and excited about your project. However, some companies may not allow the release of this type of information even with a CDA in place and any disclosures should be cleared through the appropriate company channels.

You will want to meet one-on-one with key facility personnel so that you can get a feel for their personality and working style, get answers to the various questions on the checklists for their areas of responsibility, and assess their general knowledge, experience, and qualifications. It is best if you can meet with the following people. Some of the titles may be different between laboratories, but their general functions and responsibilities will be similar:

- Facility general manager (head of the entire facility)
- Head of the Study Director group
- Study Director(s) that is likely to run your study
- Head of the in-life group
- Head of animal husbandry (if separate from the in-life group)
- Head of formulations
- Head of analytical
- Head of clinical pathology
- Head of pathology
- Veterinary pathologist (if not the head of pathology)
- Head of general operations (e.g. facility maintenance, HVAC)
- Head of report writing
- Head of computer operations and validation
- Head of QAU
- Head of archives
- Attending veterinarian
- Chair of the IACUC
- Heads of other specialized groups.

During your meetings, you can get answers to the various questions on the checklists and any other questions you may have for their area of expertise and your study. By interacting with each person, you will be able to assess each person's knowledge and their willingness to help you, which is likely a reflection of how they run their portions of studies. One of the key meetings will be with the potential Study Director for your study. You will want to assess thoroughly their experience with your type of study, knowledge, qualifications, training, attention to detail, coordination of study functions, working relationship/style, and ability to serve as the single point of control for the study. Whether the study is being conducted in accordance with GLPs or it is non-GLP, the Study Director plays a pivotal role in ensuring that all study functions are run according to the protocol and that data are collected and reported accurately. The Study Director will be your main point of contact for the study and will communicate your requirements to the rest of the laboratory. Therefore, it is essential that you can establish a good working relationship with the Study Director, they are responsive to your needs, and they run a sound study. The following are some items that can help assess the ability of the Study Director to run your study:

- How many studies have they conducted of the given study design and species?
- How many studies do they run at a given time (i.e. do they have sufficient time to oversee your study)?
- What are their qualifications and training as a Study Director?
- What is their level of attention to detail (e.g. do they raise relevant questions after reading the study outline, how do they track and stay on top of study progress)?
- Are they responsive to the needs of the Contracting Scientist (e.g. if you ask for a non-routine item to be included in the study, do they actively investigate what is needed)?
- What is the process for developing a new protocol?
- How often does the Study Director visit the animal rooms and observe various study functions (e.g. dosing, formulation preparation, necropsy)?
- Does the Study Director know the various technicians and supervisors by name (this is an indication that the Study Director is involved with the day-to-day conduct of studies)?
- How often does the Study Director audit the raw data and what does this entail?
- What is their procedure for contacting you about study deviations, moribund or dead animals, and any other study-related issues?
- How often do they provide study progress updates and what is the format (e.g. summary tables, verbal summary)?
- How do they ensure that study functions remain on track and are appropriately conducted? Are there special procedures used for functions conducted at off-site facilities?
- What is the working relationship between the Study Director and the QAU (e.g. congenial versus confrontational)?
- Does the Study Director thoroughly review the draft and final reports before release?

The last main item of your auditing entails reviewing various facility documents in order to complete the remainder of the checklist items. For example, you should audit several personnel files to ensure they have adequate documentation of training and experience and you should review the general protocol and reporting templates. One of the main reviews should be assessing the quality of the SOPs. SOPs drive many of the critical laboratory functions and it is essential that they be clear and contain all of the necessary details to complete the given procedure. You will be able to assess thoroughly the quality and accuracy of the majority of the SOPs since they will be for fairly generic procedures and processes. Some may be very specific and you may not be able to assess the technical aspects but you can still assess whether it meets the general requirements of a valid SOP (e.g. most recent version, approval by management, clearly written). Depending on the size of the laboratory, there may be many SOPs for procedures that do not apply to the study you will be running and it is probably not necessary to review those. Depending on the laboratory, there may be an excessive number of SOPs or the bare minimum. In addition, each SOP may be extremely detailed and lengthy or it may be very short and concise. Besides having SOPs that meet the basic GLP requirements, there is really no right or wrong answer as to the correct number of SOPs and the amount of detail that is needed in each one. What is critical is that SOPs exist for at least the basic GLP requirements, SOPs exist for other critical study and facility functions and procedures, and the SOPs provide enough detail and are clearly written such that a person trained in the SOP can easily understand and follow the instructions. The ultimate goal of SOPs is to ensure that study and facility functions and procedures are conducted in a consistent manner. This is important since most facilities have multiple personnel conducting the same function on a given study and across studies and consistency in conducting the function is essential to generating reliable and consistent data. Since it is essential that only the most recent versions of the SOPs are used by facility personnel, during your walk through of the facility, you need to check to make sure that outdated SOPs are not being used. You can do this by recording the version and approval date of several of the SOPs actively being used in the facility and compare that back to the master SOP file. If possible, you should also assess if personnel are following the SOPs by observing some of the procedures (e.g. are daily health checks and clinical observations being conducted according to the SOP?).

Once you are finished with your reviews, you should have a final wrap-up session with your host, general manager, and any other relevant personnel so that you can review your findings and they can provide answers or responses. You should also go over any outstanding deliverables such as study timing and pricing quotes. You should complete the laboratory selection checklist including any detailed comments about the laboratory that will help drive your decision about using them for future studies. With all of this information, you will have a sound basis for not only determining the ability of the laboratory to run your study but also for comparing multiple laboratories to each other in terms of both quality and pricing.

LABORATORY SELECTION CHECKLIST

Laboratory Selection Checklist	
Date of Visit:	
Audited By:	
Laboratory Name:	
Address:	
Telephone / Fax:	
Personnel Visited:	Title and Phone Number/e-Mail:

LABORATORY SELECTION CHECKLIST	YES	NO	Comment
Personnel			
1. Facility general manager/Institutional Official Credentials/experience			
2. Head of study director group Credentials/experience			
3. Head of the in-life group Credentials/experience			
4. Head of animal husbandry Credentials/experience			
5. Head of formulations Credentials/experience			
6. Head of analytical Credentials/experience On site or contracted?			
7. Head of clinical pathology Credentials/experience On site or contracted?			
8. Head of pathology Credentials/experience On site or contracted?			
9. Veterinary pathologist Credentials/experience On site or contracted?			
10. Biostatistician Credentials/experience On site or contracted?			
11. Head of general laboratory operations Credentials/experience			
12. Head of report writing Credentials/experience			
13. Head of computer operations and validation Credentials/experience			
14. Head of QAU and auditors Credentials/experience On site or contracted?			

LABORATORY SELECTION CHECKLIST	YES	NO	Comment
15. Head of archives Credentials/experience			

16. Attending veterinarian
    Credentials/experience
    On site or contracted?			

17. Chair of the IACUC Credentials/experience			
18. Access to specialists (e.g. Ophthalmologist, Cardiologist, etc.)			
19. Other			
20. Technicians for various functions are qualified, experienced, and trained			
21. Sufficient time and staff available to conduct the study			
Study Outline			

1. Reviewed study outline
    Can the laboratory conduct the study?
    What portions *cannot* be conducted?
    What portions will be contracted out?			

2. Facility can meet the required timelines for initiating, performing, and reporting the study			
3. Reviewed product/project background information			
Study Director			
1. Has experience with the given study design			
2. Has experience with the species			
3. Qualified and trained as a Study Director Degrees, credentials, certifications			
4. Detail oriented			
5. Responsive to Sponsor needs			
6. Routinely visits animal rooms and observes critical study functions			
7. Knows the technicians by name			
8. Routinely audits raw data What does this entail?			
9. Immediately contacts Sponsor about serious study deviations, moribund or dead animals, and other significant study-related issues			
10. Provides study progress updates What is the format and frequency			

LABORATORY SELECTION CHECKLIST	YES	NO	Comment
11. How does the Study Director ensure that all study functions are conducted appropriately and remain on track?			
12. How does the Study Director handle oversight of portions of the study that are conducted at other sites?			
13. Does the Study Director have a good working relationship with the QAU?			
14. Study Director thoroughly reviews draft and final reports before release			
Facilities and Equipment			
1. Were the following areas inspected and found to be satisfactory: Animal receiving area Animal rooms Animal veterinary treatment area/ surgery/ emergency care Cage and equipment washing and storage Feed and bedding storage General supplies storage General receiving, including test article receipt Formulations, including test article storage and mixture preparation Analytical laboratory Clinical pathology laboratory Necropsy QAU Archives Other			
2. The equipment required for the study was inspected and found to be appropriately maintained, calibrated, and/or monitored List key pieces of equipment that were inspected			
3. The laboratory facilities are constructed and organized in a manner that facilitates study conduct (e.g. cage wash is near animal rooms)			
4. The facilities are clean and organized			
5. Appropriate security procedures are used for limiting access to authorized personnel only (e.g. animal rooms)			
6. Appropriate personnel protective equipment (PPE) is used List the PPE required to enter animal rooms:			
Animals and Animal Care			
1. Animals are obtained appropriately and from quality sources			
2. Animals are used and cared for in a humane manner and in compliance with the animal welfare regulations			
3. There is adequate veterinary staff and veterinary oversight			
4. Technicians are qualified, trained, and experienced in the humane care and use of animals			
5. What type of housing is used for various species?			
6. What type of environmental enrichment is used for various species?			

LABORATORY SELECTION CHECKLIST	YES	NO	Comment
QAU and SOPs			
1. QAU is in place and independent from the conduct of the study Who does the QAU report to:			
2. SOPs in place that cover the functioning of the QAU			
3. SOPs in place that cover the range of activities needed for the study			
Data Collection			
1. How are data collected (e.g. paper, electronic, combination)?			
2. Electronic data capture systems are validated and compliant with GLPs and electronic records regulatory requirements (e.g. 21 CFR Part 11)			
Regulatory			
1. Facility is AAALAC accredited			
2. IACUC is in place How often do they meet?			
3. Results of the most recent animal welfare inspections (e.g. USDA) and/or annual reports (e.g. USDA, OLAW) were satisfactory			
4. Facility has GLP Certificate (typically for facilities outside of the USA) Who issued the certificate (e.g. EU, OECD) and when?			

5. Results of the most recent regulatory
    inspections (e.g. FDA GLP, USDA, OECD,
    etc.) were satisfactory. If not, were the
    responses to the inspections satisfactory?			

6. Has the site ever been disqualified to conduct GLP studies? If so, what
    were the reasons for disqualification and how did the laboratory address
    them?			

General Comments			
Follow-up and Deliverables			

Final Recommendation	
This site is recommended for use as a study site:	☐ **is recommended**
This site is *NOT* recommended for use as a study site:	☐ **is *NOT* recommended**
Comments on reason for decision:	
Name of auditor	
Signature of auditor	Date

Chapter 6

Project Proposal

Karen VanLare BA, RVT* and William F. Salminen PhD, DABT, PMP†
*Novartis Animal Health, Greensboro, NC, †PAREXEL International, Sarasota, FL

Key Points
- The project proposal provides all of the key details of the study
- The project proposal helps the laboratory clearly understand the study design and requirements
- The project proposal helps you obtain accurate quotes for easy comparison between laboratories

---

Once one or more laboratories are screened and selected as potential candidates for running a nonclinical study, the next step is obtaining price quotes for the study(s). This chapter focuses on obtaining accurate quotes for easy comparison between laboratories and negotiating the price. This process also applies to laboratories which you have utilized for past studies (i.e. they are already qualified) and you are now interested in placing a new set of studies at the laboratory. A study outline template is provided at the end of the chapter, which will help the Contracting Scientist convey the critical aspects of the study to the laboratory so that the laboratory can assess their ability to conduct the study and an accurate quote can be prepared. Depending on the company, certain legal documents might have to be in place (e.g. confidential disclosure agreement) before certain parts of the project proposal process are conducted and these are addressed in the next chapter. Also, as mentioned in the previous chapter, it might be helpful to provide one or more detailed study outlines during the auditing of a new laboratory so that you can more accurately assess the ability of the laboratory to conduct the study during your visit.

Laboratories often have standard pricing for routine studies that follow specific guidelines (e.g. Organisation for Economic Cooperation and Development [OECD] or International Conference on Harmonisation [ICH]). However, even with these standard studies, there are a lot of options for running the study

Nonclinical Study Contracting and Monitoring. http://dx.doi.org/10.1016/B978-0-12-397829-5.00006-5

(e.g. how many times will clinical pathology samples be collected, will pharmacokinetic samples be collected and will the laboratory conduct the analytical work) and these will all impact the final price of the study. Therefore, whether your study follows more routine guidelines or it is a very unique study, it is very helpful to fill out a detailed study outline that covers all of the major aspects of the study so that the laboratory clearly understands your needs and can generate an accurate price quote. Even with a detailed study outline, it is important to keep in mind that aspects of the study might change during the drafting of the study protocol and these changes might impact the price of the study. However, providing a detailed study outline prior to drafting the protocol will allow the laboratory to provide a very accurate quote and also assess if they can conduct the study or if certain aspects of the study may require additional support or have to be conducted at other facilities (e.g. analytical work).

DETAILED STUDY OUTLINE

This section covers completing the detailed study outline. Once completed, this can be submitted to the laboratory so that they can review it, ask any clarifying questions, and generate a price quote. A blank study outline template is provided at the end of this chapter. Examples of each major section will be covered in this section. The study example that will be used is a 2-week canine toxicology study that includes telemetry for cardiovascular function monitoring and pharmacokinetic sampling for exposure analysis. The test article (ABC-123) is a pure drug active that needs to be formulated in a suspension aid (0.5% aqueous methylcellulose) and requires analytical verification of concentration, stability, and homogeneity. The fictional company sponsoring the study is called Z-Pharm.

The first part of the study outline covers basic information about the study, the main Sponsor and laboratory contacts, and if the study needs to be conducted according to specific regulations (e.g. US Food and Drug Administration Good Laboratory Practices [US FDA GLPs]). The Sponsor Representative and Sponsor Monitor may be the same person, particularly at smaller companies where the Contracting Scientist plays both roles.

Study Title:	Two-week repeat dose toxicology study in beagle dogs using ABC-123
Study Number:	Z-Pharm #2012-001
Overview	
Introduction/Purpose:	The objective of this study is to evaluate the safety of ABC-123 when administered once daily via oral gavage to beagle dogs over a 2-week period. Pharmacokinetic sampling will be conducted at select times to verify exposure levels

Study Type:	Two-week dog toxicity study
Sponsor:	Z-Pharm, Inc. 123 Research Way Princeton, NJ 08540
Sponsor Representative:	Kathleen Smith PhD DABT Room 456 123 Research Way Princeton, NJ 08540 T: 800-555-1515 x45 E: ksmith@z-pharm.com
Sponsor Monitor:	Joe Derringer Room 57 123 Research Way Princeton, NJ 08540 T: 800-555-1515 x12 E: jderringer@z-pharm.com
Contract Laboratory:	*Name/Address:* Unitox, IN, USA *Main Contact:* Sam Jones *Role:* Sales representative *Telephone:* 888-555-2121 x103 *E-mail:* sam.jones@unitox.com
Laboratory Study Number:	To be determined
Regulatory Compliance:	GLP (US FDA and OECD)

The next part of the study outline covers information about the type of animal model and the age, weight requirements, and/or sex requirements. You can also specify if you have a preferred source for the animals (e.g. you prefer a given breeder since their animals have calm temperaments and they are acclimated to routine study procedures from a young age).

Animal Model	
Species:	Canine – purpose bred beagles
Age:	≤4 months of age on the first day of dosing
Body Weight:	Determined by age
Sex:	Equal numbers of males and females
Source:	Preferred source is Justin Research Breeders

Information on the test and control articles is covered in this section. If the test article needs to be formulated (e.g. in a carrier to create a suspension or in the feed), you need to specify the relevant information. Also, indicate if you need the laboratory to conduct analytical verification of concentration, stability, and homogeneity. You will need to indicate if an analytical method is available or if the laboratory needs to develop one from scratch. Even if a validated method is available for the given formulation, it will have to be transferred to the laboratory, which entails time and additional cost. It is best to provide any relevant analytical methods to the laboratory so that they can assess their ability to conduct the analytical work. Also, under Good Laboratory Practices (GLPs), the test article needs to be characterized. You should specify if the relevant characterization information will be provided to the laboratory or retained on file by the Sponsor.

Test and Control Articles:	
Test Article:	ABC-123 (pure active ingredient). Will be supplied by the Sponsor along with characterization documentation
Control Article:	0.5% methylcellulose (in water). Supplied and prepared by the laboratory
Formulation:	Require laboratory to prepare suspension of ABC-123 in 0.5% methylcellulose
Analytical Verification Required:	Verify concentration, stability, and homogeneity. Stability needs to be assessed over at least a one-week period prior to the start of the study. Verify concentration and homogeneity at typical intervals for a 2-week study involving weekly dose preparations
Validated Analytical Method Available:	GLP validated method is available for test article in 0.5% methylcellulose and is attached to this study outline

Basic study design information is covered in this section. Most of the items are self-explanatory, but a few deserve special mention. You should specify the type of housing since this can affect pricing and also only a single type of housing may be available for the species (e.g. many laboratories house dogs in cages and do not have large runs). For many studies, acclimation period assessments are limited; however, in this study, a detailed assessment of the baseline health of the animals is required, which is why many different assessments are conducted during the second week of the acclimation period. For the Exposure Period/Regimen, it is important to note if there are any special procedures that are needed, in this case, the dose must be followed by a flushing dose of water. The Blinding of Study Functions is included as an item since some studies require that the technicians and/or pathologist are blinded to treatment.

For this study, this is not required. At the end of the study outline, a detailed timeline of events is provided. This is another means of conveying the various study-related events so that the laboratory has a clear understanding of what is required in the study.

Basic Study Design:	(see also Table 6.1 for timeline of events)
Number of Animals:	4 dogs/sex/group. 4 treatment groups. Order at least 2 extra animals per sex to ensure only healthy animals are enrolled. Total = 18 male and 18 female dogs
Housing:	One animal per cage
Route of Administration:	Oral gavage (using a stomach tube)
Control Treatment:	Vehicle (0.5% methylcellulose)
Test Article Treatment:	Dose Groups: 10, 30, and 100 mg/kg of ABC-123. Rationale for Selection: A single dose range finding study showed that 100 mg/kg caused minimal adverse clinical signs and minimal focal hepatocyte degeneration via histopathological analysis
Acclimation Period:	At least 2 weeks. The following assessments are needed during the 2nd week of acclimation: (1) baseline clinical pathology samples, (2) physical examination, (3) ophthalmology examination, (4) baseline cardiovascular assessment using telemetry (LifeShirts), (5) daily feed consumption, and (6) daily body weight
Exposure Period/Regimen:	Once daily oral gavage using a 1 ml/kg dose volume for 2 weeks. Each dose is to be followed by 1 ml/kg of water to wash the dose solution from the tube. Dosing must be conducted in the morning at approximately the same time each day
Fasting:	In order to increase the bioavailability of the test article, animals must be fasted starting at ≈4–5 p.m. each night until the dose the next morning. Feed can be offered to the animals 2 h after dosing
Randomization:	Animals must be weight ranked and then randomly assigned to dose groups. Mean weight must not be >1 kg between groups. Separate randomizations for males and females
Blinding of Study Functions:	Not required

TABLE 6.1 Summary of Dose Groups and Timeline of Events

Two-week repeat dose toxicology study in beagle dogs using ABC-123

Study Number: Z-Pharm #2012-001

Treatment Groups

Group	Test Article	Dose (mg/kg)	# Animals
1	0.5% MC	NA	4 m / 4 f
2	ABC-123	10	4 m / 4 f
3	ABC-123	30	4 m / 4 f
4	ABC-123	100	4 m / 4 f

Study Timeline

Study day	Fasting	Dose	BW	GH/M/M	CO	FC	PE	Blood	Urine	CV	PK	Necropsy	Other
−14 to −8				X									
−7			X	X		X							
−6			X	X		X							
−5			X	X		X	X[1]	X[1]	X[1]	X			Eye[1]
−4			X	X		X							
−3			X	X		X							
−2			X	X		X							
−1	X		X	X		X							

									Eye
1	X	X	X	X	Multiple	X	X		Multiple	
2	X	X	X	X	XX	X	X^2	X	X^3	
3	X	X	X	X	XX	X				
4	X	X	X	X	XX	X				
5	X	X	X	X	XX	X				
6	X	X	X	X	XX	X				
7	X	X	X	X	Multiple	X	X		Multiple	
8	X	X	X	X	XX	X	X^2	X	X^3	
9	X	X	X	X	XX	X				
10	X	X	X	X	XX	X				
11	X	X	X	X	XX	X				
12	X	X	X	X	XX	X				
13	X	X	X	X	XX	X				
14	X	X	X	X	XX	X	X			Multiple
15		X	X	X	XX	X	X^2	X	X^3	X

BW: body weight; GH: general health check; CO: clinical observation; FC: feed consumption; PE: physical examination; CV: cardiovascular assessment; PK: pharmacokinetic sampling

[1] Physical and ophthalmology examinations conducted and blood and urine collected one day during the 2nd week of acclimation.
[2] Blood and urine collected prior to the daily dosing.
[3] Blood sample collected prior to the daily dosing.

The different endpoints that will be assessed in the study are covered in this section. Different laboratories have different terminologies for some of the assessments, so it is important to provide sufficient details about each one and the timing so that they can match up their procedures to your requirements. For example, in this study, clinical observations involve a thorough assessment of the health of each animal and this might be beyond what the laboratory normally considers a routine clinical observation. In addition, on select days, these observations must be conducted at frequent intervals in order to capture thoroughly potential effects caused by the test article. For the clinical pathology endpoints, the laboratory may ask you for a list of the specific tests you require so that they can assess their ability to run the different tests (e.g. if there are tests that they do not have validated) and can price the testing accordingly. In this study, electrocardiograms and other cardiovascular assessments are being collected using non-invasive LifeShirts, which are worn by the animals and the signals captured by telemetry. This should be clearly indicated so that the laboratory can determine their ability to provide this service or provide alternatives to the assessment. For the eye exams and necropsy/histopathology, board-certified veterinarians are required. It is important to indicate this clearly since some laboratories may have a regular veterinarian conduct the eye exams and/or oversee the necropsy. The laboratory may also ask you for the specific list of organs to be weighed and collected and read for histopathology.

Variables:	
General Health Checks:	Twice daily (a.m. and p.m.) for general health, moribundity, mortality, and access to feed and water. Observations to start at start of acclimation period
Clinical Observations:	Regular Clinical Observations: Twice daily (a.m. and p.m. at least six hours apart). Observations will include, but not be limited to eyes, integument, mucus membranes, respiratory system, circulatory system, autonomic and central nervous systems, somatomotor activity, behavior pattern, and gastrointestinal tract (e.g. vomiting, regurgitation and stool characteristics). Particular attention will be directed to observations of seizures, tremors, salivation, vomiting and diarrhea (these effects were observed in the range finding study) Special Clinical Observations: On study days 1 (first dose) and 7, frequent clinical observations need to be made at the following timepoints: immediately pre- and post-dose and at 1, 2, 3, 4, 5, 6, 8, 10, 12, and 18 hours
Physical Examinations:	Detailed physical examination conducted by a veterinarian once during acclimation (2nd week of acclimation) and then on study days 1, 7, and 14 (≈ 6 h post-dose on each day)

Adverse Reactions:	If adverse reactions are observed, the observation period will be extended to a time at which no further adverse reactions are observed. This would include time of death or euthanasia. An attempt will be made to collect blood and urine for clinical pathology for any animal where euthanasia appears imminent
Body Weight:	Daily starting the 2nd week of acclimation. Final terminal body weight prior to necropsy. Dosages adjusted daily based on the daily body weight collected prior to dosing
Feed Consumption:	Daily starting the 2nd week of acclimation
Water Consumption:	Not determined
Clinical Pathology:	Once during the 2nd week of acclimation (baseline), on study days 2 and 8 (blood collected prior to the daily dose), and then prior to necropsy on study day 15. Blood for routine hematology and clinical chemistry panels. Urine collected from cage pan for urinalysis
Electrocardiogram:	Cardiovascular assessments conducted using telemetry (using LifeShirts) once during the 2nd week of acclimation (baseline) and on study days 2, 8, and 14
Ophthalmology Exams:	Once during the 2nd week of acclimation and on study day 14. Must be conducted by a board-certified veterinary ophthalmologist
Necropsy:	Conducted on study day 15 approximately 24 h after the last dose. Necropsy order must cycle one animal at a time through each dose group in order to minimize temporal bias. Animals are to be fasted prior to necropsy. Gross necropsy is to be overseen by a board-certified veterinary pathologist
Organ Weights:	All routine organs
Histopathology:	All routine organs. All dose groups are to be processed and read by a board-certified veterinary pathologist
Other:	Blood collected from a peripheral vein for pharmacokinetics on study days 1, 7, and 14. Blood collected ≈1 h pre-dose (first dose only for baseline) and at 1, 4, 8, and 24 h post-dose on each day. Blood collected in serum separator tube and serum frozen at $-80\,°C$ until shipment to Sponsor for analysis. Sponsor will analyze samples using validated method under GLPs and will provide sub-report for inclusion in the study report

The last two sections cover information about your requirements for starting the study and reporting the results. This information will help the laboratory determine if they can meet your timelines and reporting requirements. For the Reporting section, a specific item regarding Raw Data is included since some regulatory submissions require copies of all raw data generated during the study; however, this is an exception and not a rule for most studies.

Study Start:	
Test Article:	Available for shipment
Analytical Methods:	Available for test article in vehicle but needs to be transferred to laboratory
Targeted Start Date:	Less than 4 months from today
Reporting:	
Raw Data:	Copies are NOT required to be included in the final report
Individual Data:	Survival, clinical observations, body weight, feed consumption, clinical pathology, cardiovascular assessments, and necropsy findings, including histopathology results
Group Summary Data:	Clinical observations, body weight, feed consumption, clinical pathology, cardiovascular assessments, organ weights (absolute and relative to body and brain weight), and histopathology results
Statistical Analysis:	Body weight, feed consumption, clinical pathology, cardiovascular assessments, and organ weights
Other:	Copies of water quality assessments, summaries of animal information (e.g. birth dates, vaccinations)
Report Format:	Standard laboratory format is acceptable
Reporting Timeline:	Draft report provided 6 months after the last dose is administered. Final report provided 1 month after comments on the draft report are provided to the laboratory
Final Report Deliverables:	Two paper copies and an electronic copy (PDF format)

A table of the different dose groups and a timeline of specific study events follows the study outline. This helps the laboratory understand the structure of the study and timing of all of the different study functions.

PRICE NEGOTIATION

After you have provided detailed study outlines to several laboratories (two to four laboratories that you have pre-screened for capabilities is typically sufficient), it may take a while for them to review the outlines, assess their capabilities for conducting all the various parts of the study, and then price the study. Each laboratory has different methods for pricing studies and it is not uncommon for there to be significant differences in pricing between laboratories, especially for larger, more complex studies. If large differences exist, you should carefully review the price quotes provided by each laboratory to ensure that they fully understood all of the key aspects of the study and what aspects are driving the price differences. It may be that a laboratory provided a low price since they did not understand what is involved with a complex procedure or perhaps the high priced laboratory included more detailed assessments than required. The most critical aspect is that you have price quotes that are reflective of what you need done during the study. If the price quote is for a study design that does not match the detailed study outline, the quote is inaccurate and the price is likely to change once the protocol is drafted. Also, having inaccurate price quotes does not allow a fair comparison between laboratories.

Significant price differences between laboratories can occur for a variety of reasons, with one of them already being mentioned (i.e. the laboratory not fully understanding the scope of the study). If a laboratory is trying to get your initial business, they may purposely underbid the study with the hopes that they can capture future work at regular prices. The laboratory might provide a bid that is so low that they actually lose money on the study just in the hope that you will place larger packages of studies in the future. You can get a nice price break in this situation; however, you should realize that future studies are unlikely to be at cut-rate prices. Some laboratories have stronger reputations in terms of regulatory compliance and scientific conduct of nonclinical studies and these laboratories will, in general, price a given study at a higher rate. There is a lot to be said about laboratories that have stellar reputations and it can be worth the extra money to pay for this quality. However, it is important to realize that each study is its own entity and it is how the laboratory conducts *your* study that is important. A laboratory that is not as well known may pay a lot more attention to your study than a well-known laboratory. This is where conducting a thorough audit of each laboratory is important. In addition, you should ensure that you can establish a solid working relationship with the Study Director since they are your main conduit for running a sound study. Some studies may be too complex or outside of the area of expertise of the laboratory. They may bid on the study but they may inflate the cost of the study since they may not really want to conduct the study and develop the necessary expertise unless they can get a premium price. The current and predicted capacity of the laboratory can also drive pricing. If the laboratory is projected to be filled for the foreseeable future, they are more likely to provide a higher price since they do not need your study to maintain operations; whereas, they may provide a cut-rate price if they are looking to fill empty animal rooms and keep their employees busy. Package pricing or history of placing work can play a role in pricing. If you plan on placing a package of studies with the laboratory, you are likely to get price breaks on the studies. Also, if you have a long-standing relationship with the laboratory and place many studies with them, they are likely to want to keep you as a loyal customer and provide you with pricing that is likely to retain your business. A final note is that some laboratories may provide quotes that are on the high side with the expectation that they will always negotiate the study at a lower price; whereas, other laboratories will put their best foot forward at the start and will not negotiate significantly.

Understanding the reasons for any pricing differences between laboratories is important for comparing prices and negotiating the final price of the study. Once you are confident that the quotes accurately reflect what needs to be done on the study, you should weigh out the pros and cons of placing the study at each laboratory. A few items to consider are:

- Price
- Regulatory compliance
- Scientific quality and expertise
- Working relationship with the Study Director
- Study start availability.

If all of the factors tip in favor of one laboratory, then it may only be necessary to negotiate the price with that single laboratory; however, most likely, the different factors will favor different laboratories and you need to decide which are most important. If you have to have a given scientific expertise, this may rule out placing the study with any other laboratory, regardless of the price. Once you have narrowed the list of laboratories, you can approach the laboratories and ask for their best pricing. Each person will have their own method and personality for the negotiations and no one method will be advocated. What is important is that you always ask the laboratory if they can lower the price and, if not, the reason for not doing so. Once you are reassured that you have the lowest price in hand, then you can make your final comparison between laboratories and make your final selection.

DETAILED STUDY OUTLINE TEMPLATE

The following outline can be used to provide the laboratory with relevant details about your study. Alternatively, the laboratory may have a template they can provide you with. The advantage to this is that the laboratory will likely provide you with a protocol template, or something similar, which can help with eventual drafting of the protocol, if you select that laboratory. The downside is that you may have to fill out different templates for each laboratory.

Study Title:	
Study Number:	
Overview	
Introduction/Purpose:	
Study Type:	
Sponsor:	
Sponsor Representative:	
Sponsor Monitor:	
Contract Laboratory:	*Name/Address:* *Main Contact:* *Role:* *Telephone:* *E-mail:*
Laboratory Study Number:	
Regulatory Compliance:	
Animal Model	
Species:	
Age:	
Body Weight:	
Sex:	
Source:	
Test and Control Articles:	
Test Article:	
Control Article:	
Formulation:	
Analytical Verification Required:	
Validated Analytical Method Available:	
Basic Study Design:	
Number of Animals:	

Housing:	
Route of Administration:	
Control Treatment:	
Test Article Treatment:	Dose Groups: Rationale for Selection:
Acclimation Period:	
Exposure Period / Regimen:	
Fasting:	
Randomization:	
Blinding of Study Functions:	
Variables:	
General Health Checks:	
Clinical Observations:	
Physical Examinations:	
Adverse Reactions:	
Body Weight:	
Feed Consumption:	
Water Consumption:	
Clinical Pathology:	
Electrocardiogram:	
Ophthalmology Exams:	
Necropsy:	
Organ Weights:	
Histopathology:	
Other:	
Study Start:	
Test Article:	
Analytical Methods:	
Targeted Start Date:	

Reporting:	
Raw Data:	
Individual Data:	
Group Summary Data:	
Statistical Analysis:	
Other:	
Report Format / Guidelines:	
Reporting Timeline:	
Final Report Deliverables:	

Chapter 7

Contracts and Business Ethics

Karen VanLare BA, RVT* and William F. Salminen PhD, DABT, PMP†

*Novartis Animal Health, Greensboro, NC, †PAREXEL International, Sarasota, FL

Key Points

- A confidentiality agreement should be established before disclosing confidential information with a laboratory
- A study contract spells out all of the key legal details, including obligations of the laboratory for conducting and reporting the study
- The Sponsor's legal or ethics department should be consulted before accepting gifts from laboratories since it could represent a conflict of interest

---

This chapter deals with the legal aspects of contracting a study that a Contracting Scientist should be aware of. This chapter does not provide legal advice on what documents are needed and how they should be worded since formal legal advice is needed for each specific situation. Examples of the types of documents that are typically used in contracting a study will be presented so that the reader has an understanding of what the various documents encompass and how they are used. In addition, business ethics will be touched upon since this often plays a role in ongoing and future contract negotiations and many companies have rules about acceptable interactions between the laboratory and Contracting Scientist.

CONFIDENTIALITY

Data confidentiality is a major concern when developing a new drug or product. Companies go to great lengths to ensure their new products are protected and the ideas cannot be stolen by other parties. There are a variety of methods that companies use to maintain confidentiality and the ultimate protection is typically a patent on the new product. However, patent protection is for a limited period and often companies will conduct studies on a new product before submitting a formal patent application so that the patent covers the new product for as long as possible. Since patent protection may not be in place when you run a study, you need another means to ensure that information about the new product is not obtained and used by competitors. This is important since you will have to disclose some important information about your product to the laboratory that will be running your study. If you do not have appropriate confidentially protection in place, it is possible that this information could end up in the hands of a competitor. Although most contract laboratories understand the importance of product confidentiality and go to great lengths to maintain securely all Sponsor-related information, it is important to establish a legally binding agreement between the Sponsor and the laboratory that is enforceable if a breach of confidentiality occurs. This is especially important for contract laboratories that run studies for many different Sponsors where the possibility of breach of confidentiality increases (e.g. a Study Director accidentally provides test article information and raw data to a competitor during an auditing visit or test article information is in plain sight when a competitor is walking through the animal facility).

Before a formal contract is in place, product information is often shared with the laboratory. Before any confidential information is shared, a Confidential Disclosure Agreement (CDA) should be established with the laboratory. This document may go by other names depending on the company, but the basic purpose is to ensure that the laboratory does not disclose any confidential information you discuss or provide to them without your specific permission. An example CDA is provided at the end of the chapter. Although a CDA should always be reviewed and approved by a legal expert, the Contracting Scientist should not only understand the terms of the CDA but should also help complete the CDA since they are in the best position to understand what the CDA should cover. CDAs involve a lot of legal terminology; however, some basic aspects about what the CDA is specifically covering require input from the Contracting Scientist since they are the ones most familiar with the project. The legal expert can then refine the wording and solicit further input to clarify the terms of the CDA.

The CDA can be a one-way or two-way agreement. The one-way is typically used if you will be sharing confidential information with the laboratory but they will not be sharing any confidential information with you. The two-way is used if confidential information is shared both ways. The determination of which type to use should be made by a legal professional for the specific situation. In the example CDA, there are notes in the different sections that need information inserted about the specific project. This is information that should initially be filled in by the Contracting Scientist since they are most familiar with the project and what the CDA should cover. Once a CDA is in place, you can then share confidential information with the laboratory. However, it is important to read carefully and understand the CDA so you know how and when you must specify to the laboratory that a given piece of information is confidential and subject to the CDA. Fortunately, most laboratories routinely deal with confidential information from multiple clients and they often have procedures in place to ensure information is not accidentally disclosed to other parties, even when the data are not specifically covered by a CDA. However, to be on the safe side, it is best to ensure a CDA is in place and clearly mark any confidential information so that the laboratory understands its obligations.

There is a note in the CDA template about possibly needing a Material Transfer Agreement (MTA) if research materials (e.g. chemicals, tissues, blood) are shared between the company and the laboratory. For most contracted non-clinical studies, MTAs are not used and instead the specific study contract or an overall Master Laboratory Agreement covers the exchange and ownership of research materials, such as the test article and other samples generated during the study. Typically, all materials sent to the laboratory and derived from a contracted nonclinical study (e.g. tissues, slides, data) are owned by the sponsoring company and this is clearly specified in the study contract or Master Laboratory Agreement. An MTA might be needed if the sponsoring company wants to share samples generated from the study with a separate institution for basic research purposes, such as an academic laboratory.

CONTRACTS

Once you have selected a laboratory to run your study, you will need to develop a contract to cover the obligations of both parties including specific deliverables and the timing of the deliverables. Contracts can be simple verbal agreements to complex legal documents. The type of contract that governs your study must meet the requirements of your company and the laboratory. Many laboratories have standard contract language that is a good starting point for negotiations; however, your company may have a preferred template to start the contract process. You should check with your legal expert on the preferred method for starting contract negotiations. A typical contract for a nonclinical study is complex since it has to cover many different legal obligations and clearly spell out what will be done and what happens if the deliverables are not provided and the timelines are not met. Although the contract is often complex and involves a lot of legal language, it is important for the Contracting Scientist to understand the basic aspects of the contract. This will ensure that the laboratory will perform the study in a manner that the Contracting Scientist expects.

At the very beginning of the contract negotiation, the Contracting Scientist should review the draft contract to see if it includes key items that will ensure that the laboratory meets its obligations. A few key items that the contract should address are:

- Payment schedule
- Confidentiality
- Ownership of study materials (e.g. test article, tissues, animals, and slides) and data
- Use of data generated from the study

- Study conduct and reporting
- Specific deliverables
- Timing of the deliverables
- Compliance to regulations
- Right to audit and site visits
- Animal welfare (3Rs, will follow Care and Use Guide, etc.)
- Intellectual property rights (ideas, designs, concepts, techniques, compounds, inventions, etc.)
- CRO acts as an independent contractor (Sponsor is not bound by the acts or conduct of the CRO)
- Insurance
- Remedies/indemnities (CRO will not hold Sponsor responsible for any liability, damages or expenses)
- Force majeure
- Use of name (CRO will obtain written permission to use the Sponsor's company name)
- Choice of Law (which state the laws in the agreement will apply)
- Non-solicitation (neither party will solicit the other for employment)
- Entire agreement; Modification (contract may not be modified without the permission of both parties)
- Addendums.

The study price and payment schedule should be clearly listed in the contract. Short-term studies will often need a large up-front payment with one or two subsequent payments to complete the study. Contracts for long-term studies, especially ones that are a year or longer, will have many payments at defined intervals (e.g. quarterly) over the course of the study. Regardless of the study duration, you should always have a final payment contingent on delivery of an acceptable final report. This is a good way to ensure that the laboratory maintains incentive to complete the study and provide a timely final report.

The contract should have language that covers confidentiality and provide similar protection as a CDA. The contract should clearly state that all study materials and data generated during the study are the ownership of the sponsoring company since they are paying for the study. The contract may also include specific language about the laboratory not being able to use any findings from the studies in patents, publications, or similar applications without specific approval of the sponsoring company.

A key component of the contract is specifying what the laboratory will do. The contract will often reference the study protocol as being the document that specifies what the laboratory will do during the study. This is why it is critical to have a solid protocol that clearly specifies what will be done during the study. Specific aspects of protocol development will be discussed in the next chapter. It is also important to understand that since the contract references the protocol, the laboratory is under a legal obligation to conduct the study according to the protocol but not anything beyond the protocol. Therefore, if you expected an additional clinical observation to be conducted during a study (e.g. you casually mentioned to the Study Director that it would be nice to add an additional clinical observation) but this was not listed in the protocol, do not be surprised if the laboratory does not conduct the clinical observation unless you amend the protocol, which would amend the contract. Conducting the additional clinical observation without amending the protocol would also be a Good Laboratory Practice (GLP) deviation, which was addressed in Chapter 2.

The contract should specify the timing of the study, particularly when it will start. This is often contingent upon providing test article to the laboratory and other items the laboratory may need to initiate the study (e.g. analytical methods for test article formulation analysis or pharmacokinetic analysis, test article characterization documentation). Having a specified starting date is beneficial since this allows the laboratory to reserve a room to ensure that space will be available to house your animals and start the study. Without a specified start date, your study may be delayed if rooms are filled up with other studies. The downside to specifying a date is that if you do not provide deliverables to the laboratory on time (e.g. test article), you may be subject to study delay penalties. Whether a laboratory chooses to exercise this option depends on the language in the contract and how eager they are to retain your business. If the laboratory is struggling to fill its capacity, it is less likely that they will ask for compensation for a study delay.

The protocol should provide details about the reporting of the study, content of the report, and delivery of a final report. However, often the specific timing of the different reporting phases is not listed in the protocol. If this is the case, it is critical that the contract clearly specifies the timing of various key deliverables. The protocol should clearly list the start date of the study, the timing of the first and subsequent doses, and the date of the necropsy. Once these are specified, the timing of the in-life portion of the study through to the necropsy is dependent on the study design and will not vary unless the protocol (and contract) is amended. Since the last necropsy date is fixed and can be easily determined, it is best to base the timing of reporting deliverables relative to the last necropsy date. Some studies will have staggered or multiple necropsies during the study. In these cases, the date of the last necropsy is often used. Once you know the last necropsy date, you can specify the delivery of a draft report relative to that date (e.g. provision of a draft report four months after the last necropsy). If you use another reference, such as the completion of the reading of the histopathology slides, you risk having an unknown draft reporting date since the duration of the reading of the histopathology slides is a variable that is often not specified and can vary widely depending on the laboratory's and pathologist's workload.

It is often useful if the protocol or contract clearly specifies the contents of the draft report. As mentioned in previous chapters, some laboratories may push a partially-completed draft report out the door just to meet the contract dead-line. If the protocol or contract does not clearly specify that a complete draft report must be provided and list the various components, then the contents of the draft report are up to the laboratory. Although from a reputation perspective, a laboratory would want to avoid sending out an incomplete draft report (e.g. key sections omitted or left as "to be determined", contributing scientist reports missing), they may feel that any potential monetary penalties are worth more than the ding on their reputation. The contract should also specify the timing of the final report. Since the draft report must be reviewed by the Contracting Scientist, it is often best if the timing of the final report is relative to the provision of comments to the laboratory (e.g. a final report provided six weeks after comments on the draft report are received by the laboratory). You may also consider including language in the contract that specifies that a final draft report incorporating the Contracting Scientist's comments will be provided for final review before finalization of the report. Many contracts will include language to protect the laboratory and ensure that they can issue a final report in a timely manner. For example, if the Contracting Scientist does not provide comments on the draft report within six months, the laboratory has the right to finalize the study and issue a final report.

A contract often covers an individual study. It can be tedious drafting, reviewing, and finalizing contracts for many studies. That is where a Master Laboratory Agreement (MLA) is useful if you routinely place studies at a given laboratory. An MLA is an overarching agreement that specifies contract details for all studies placed by the sponsoring company at the laboratory. The MLA covers confidentiality terms and many other contract details. Each new study placed at the laboratory is then added as an addendum to the MLA so that the study is covered by all of the legal bindings listed in the MLA. As with any contract, the MLA should be carefully established and thoroughly reviewed by a legal expert so that the company's interests are fully covered. Many of the items discussed above for individual contracts also apply to the MLA. The Contracting Scientist should play an active role in providing input and reviewing the MLA since all of their studies placed at the given laboratory will be covered by the terms of the MLA.

MAINTAINING CONFIDENTIALITY DURING THE STUDY

Once you have a contract in place and are going to start your study, you should ensure that confidentiality is maintained throughout the study. As mentioned above, laboratories often have procedures in place to prevent other clients from viewing confidential information; however, they are not fool proof. You may want to consider coding of the test article to ensure that even if a competitor accesses your study documentation, they will not be able to determine what is actually being tested unless they have the cross-reference information. Study numbers should also be coded in a manner that does not identify the sponsoring company (e.g. study # Z-Pharm-123 would not be a good code). Laboratories will often use a fixed pre-fix or designation for a given Sponsor (e.g. study #123-456 with the 123 pre-fix representing the company Z-Pharm and used for all Z-Pharm studies) and this is acceptable so long as the sponsoring company cannot be easily identified from the study number. A potential concern is when a competitor might accidentally observe paperwork or information from one of your studies that is being conducted. By using codes, a competitor who is walking through the facility (e.g. during an audit of their study) might be able to determine the type of study that is being run but would have a hard time identifying the actual test article and sponsoring company.

If your project and accompanying studies require strict confidentiality, it is best to make this very clear to the Study Director and Test Facility Management. The Study Director and Test Facility Management will then ensure that all study personnel understand that under no conditions should study details or data be discussed or shared with anyone who does not have authorization, even within the laboratory. Restriction of communication is a good way to ensure that the minimum number of people have access to your confidential information. When this is combined with coding of a test article, even study personnel within the laboratory will have a hard time determining what is actually being tested, which enhances the ability to maintain confidentiality. If the test article is coded, the Study Director and Test Facility Management will still need to be provided with the test article information so that they can ensure appropriate personnel protective equipment is used and can cross-reference the coded test article back to the actual test article in the study report. Also, other select laboratory personnel may need to know the actual test article identification (e.g. technicians conducting test article formulations analysis).

Another area that could compromise confidentiality is the transmission of information via the Internet and e-mail. There are various means to send study-related communications and data using encryption and password protection. It is best to check with your information technology experts on the best method for secure transmission. In addition, many laboratories now have secure portals where you can obtain updates on your study and transfer documents. This is a great way to transmit securely confidential information; however, you may want to have your information technology experts confirm that the laboratory's portal and means of transmission are truly secure before you start transmitting confidential documents and data. Also, some of these portals may not be compatible with company firewalls or may require specific settings to allow their use.

BUSINESS ETHICS

As mentioned in previous chapters, Contract Research Organizations (CROs) are in the business to make money by retaining existing clients, attracting new clients, and keeping animal rooms filled with studies. As with any business that seeks to retain existing business and attract new business, they must actively market themselves. Some CROs use minimal marketing and rely heavily on their reputation and word-of-mouth; whereas, others spend significant resources on selling their services and actively seeking out new clients. The authors have used CROs that span the horizon on their marketing approaches. You should not be swayed by a fancy sales pitch or be hesitant to use a CRO with a lower key sales approach. The key is to determine clearly if the CRO can provide the services you need and then conduct a thorough audit of the laboratory as described in the previous chapters. Only then can you make an accurate assessment of the laboratory's capabilities and quality.

As part of doing business and keeping the client happy, many CROs will provide small gifts during visits (e.g. coffee mugs and other items with the CRO's logo) and take you out to lunch or dinner. Most Sponsor companies allow the CRO to give small gifts and pay for meals. However, it is best to check with your company before accepting any free items and meals in case it goes against company policy. For example, when the US Federal Government contracts a nonclinical study with a CRO, federal employees who audit the laboratory/study may not accept any free gifts or meals since that would be considered a conflict of interest. More and more private companies are enacting similar restrictions. Regardless of company policy, extravagant gifts should always be avoided since these present a clear conflict of interest. If your management discovered that you received a large gift before placing a study with a CRO, it would appear that the CRO "bought" your decision to place the study at that laboratory. This even applies to gifts given after the study has been awarded, especially if you plan on placing future studies with the CRO. Two examples highlight extravagant offers that the authors have had to turn down. One CRO offered to pay for a weekend's stay at a nice hotel, meals, and attend a NASCAR event with seats close to the racecourse (i.e. very exclusive and expensive). The offer was turned down since not only was this an extravagant gift, but the author's company specifically prohibited such gifts. In addition, the CRO made the offer shortly after being asked to bid on a large package of studies. The second example involved a CRO that offered to use their corporate jet to fly one of the authors from the laboratory to the nearest large city to watch a professional baseball game. This offer was extended after several large studies were placed at the CRO; however, it was clear that this was a means to try and retain business. The offer was declined.

EXAMPLE OF A CONFIDENTIAL DISCLOSURE AGREEMENT

CONFIDENTIAL INFORMATION DISCLOSURE AGREEMENT

Effective Date:

1. The parties to this Agreement are:

Z-Pharm, Inc.

123 Research Way

Princeton, NJ 08540

2. The parties appoint the following representatives to disclose and receive "CONFIDENTIAL INFORMATION" which is more particularly described below.

For Z-PHARM	For PARTICIPANT
Kathleen Smith PhD DABT Room 456 123 Research Way Princeton, NJ 08540 T: 800-555-1515 x45 E: ksmith@z-pharm.com	

3. WHEREAS Z-PHARM and PARTICIPANT respectively possess certain confidential and proprietary information and are willing to disclose such confidential and proprietary information to each other, subject to the terms and conditions set forth herein.

4. Information disclosed to be used solely for the purpose of evaluating _____ *(NOTE: INSERT A BRIEF DESCRIPTION OF THE WORK [e.g. the toxicity of ABC-123 in various nonclinical species])*_____. The Parties shall disclose as follows:

Z-PHARM shall disclose to PARTICIPANT certain confidential and proprietary information and materials relating to: (a) _____.

[NOTE: be as general as possible in describing the nature of the Z-PHARM confidential information so that the confidentiality obligations will be broad

enough to cover all information or materials intended to be disclosed or that may be disclosed.]

And PARTICIPANT shall disclose to Z-PHARM certain confidential and pro-prietary information and materials relating to _____.

[NOTES: if possible, define the confidential information that Z-PHARM receives more specifically so as to be clear as to Z-PHARM'S obligations. Delete this Participant section if only Z-PHARM information is disclosed (i.e. a one-way agreement). A Materials Transfer Agreement may be required if chemical or biological materials (e.g. chemicals, tissue samples, etc.) are being transferred.]

5. A recipient of CONFIDENTIAL INFORMATION disclosed under this Agreement shall not use said information except for the purpose of evalu-ating the CONFIDENTIAL INFORMATION for Z-PHARM and reporting to Z-PHARM on any applications of the CONFIDENTIAL INFORMATION in the fields of animal or human health.

6. CONFIDENTIAL INFORMATION may include, by way of example but without limitation, data, know-how, concepts, proposals, formulae, processes, designs, sketches, photographs, plans, drawings, specifications, samples, reports, cus-tomer lists, pricing information, studies, findings, inventions and ideas. To the extent practical, CONFIDENTIAL INFORMATION shall be disclosed in documentary or tangible form marked "CONFIDENTIAL" or "PROPRIETARY," but the failure to do so shall not nullify the confidential or proprietary nature of the disclosure. In the case of disclosures in non-documentary form made orally or by visual inspection, the discloser shall identify such Information as CONFIDENTIAL at the time of disclosure and shall have the right, or if requested by the recipient, the obligation to confirm in writing the fact and general nature of each disclosure within a reasonable time after it is made. Each representative designated by the parties for receiving and disclosing CONFIDENTIAL INFORMATION shall make all arrangements for his/her party to be informed of all communications relating to this Agreement. (The amount of CONFIDENTIAL INFORMATION to be disclosed is completely within the discretion of the discloser.) The recipient of CONFIDENTIAL INFORMATION shall retain such in confidence and prevent its disclosure to any third party, in each case using the same degree of care as it uses to protect its own confi-dential information, but not less than reasonable care, and shall limit internal dissemination of CONFIDENTIAL INFORMATION within its own organization to individuals whose duties justify the need to know such information, and then only provided that there is a clear understanding by such individuals of their obligation to maintain the trade secret status of such information and to restrict its use solely to the purpose specified herein.

7. The CONFIDENTIAL INFORMATION will be disclosed by: (circle as appropriate)

Z-PHARM   PARTICIPANT   Both Parties

8. Recipient agrees not to chemically analyze or have analyzed the sample product(s) unless written permission is obtained from Z-PHARM.

9. The fact of discussions between Z-PHARM and PARTICIPANT, as well as the contents of such discussions, are also considered CONFIDENTIAL INFORMATION hereunder.

10. A recipient shall be under obligation for (10) ten years from the Effective Date above and shall not use CONFIDENTIAL INFORMATION for a purpose other

than that for which it was disclosed and shall not disclose the CONFIDENTIAL INFORMATION to any third party and shall use the same precautions to prevent disclosure of CONFIDENTIAL INFORMATION to third parties as the recipient uses to prevent the disclosure of its trade secrets to third parties.

11. This Agreement shall not prevent Z-PHARM from disclosing CONFIDENTIAL INFORMATION to employees of its other affiliated corporations so long as the employees of any such affiliates receiving such CONFIDENTIAL INFORMATION agree to observe the obligations of this Agreement.

12. The PARTICIPANT agrees that all information, discoveries, inventions, improvements or data collected, generated, or prepared during this agreement, whether patentable or not shall be the property of Z-PHARM. The PARTICIPANT represents that each of its employees has agreed to assign to the PARTICIPANT all inventions made by such employee in the course of his or her employment. The PARTICIPANT agrees that if, during the term of this agreement, any of its employees shall make an invention which related directly to the articles, concepts, or substances studied, the PARTICIPANT will promptly make the invention known to Z-PHARM. At the request of Z-PHARM, the PARTICIPANT agrees to assign to Z-PHARM any and all rights, titles, and interest to the invention. The PARTICIPANT shall upon request assist Z-PHARM in connection with the preparation and prosecution of any application for letters patent or certificate of invention relating to the invention. All reasonable expenses incurred by the PARTICIPANT and associated with establishing Z-PHARM's patent rights shall be paid by Z-PHARM.

13. At the discloser's request, the recipient will return or destroy all CONFIDENTIAL INFORMATION in its possession or under its control (including all copies thereof, all unused samples and all documentation embodying CONFIDENTIAL INFORMATION and all CONFIDENTIAL INFORMATION in computerized form, whether on disks, hard drives, source codes, object codes or otherwise) to discloser. In the case of destruction, recipient will provide a certificate signed by an officer of recipient, certifying to the destruction of all CONFIDENTIAL INFORMATION aforementioned.

14. PARTICIPANT warrants that they are free from any commitments that would restrict PARTICIPANT in fulfilling their obligations under this Agreement.

15. No other right to CONFIDENTIAL INFORMATION is granted hereby, and nothing contained in this Agreement shall be construed as creating an express or implied license to practice CONFIDENTIAL INFORMATION.

16. In the event that any Government Authority (as hereinafter defined) requires or requests the disclosure of any or all of Z-PHARM'S CONFIDENTIAL INFORMATION, Participant will give prompt and timely notice to Z-PHARM of the required or requested disclosure, so as to give Z-PHARM the opportunity of contesting such disclosure. In addition, Participant will use its best efforts to make a claim of confidentiality with respect to such CONFIDENTIAL INFORMATION (including, without limitation, providing all INFORMATION necessary to support such a claim) and will otherwise cooperate with Z-PHARM in responding to the request or requirement of such Governmental Authority. Z-PHARM will pay all reasonable costs incurred by Participant in such response. For purposes of this agreement "Governmental Authority" means any federal, state or local government body or agency (including any

executive, legislative or judicial body, court of law or independent agency) which exercises authority or control over a party of the business conducted by a party, and includes all orders, summons, subpoenas and like demands issued or at the direction of such Governmental Authority.

17. The recipient of CONFIDENTIAL INFORMATION shall be under no obligation with respect to any information: (a) which is, at the time of disclosure, available to the general public; or (b) which following disclosure becomes available to the general public through no fault of the recipient; or (c) which recipient can demonstrate was in its possession before receipt; or (d) which is disclosed to recipient without restriction on disclosure; or (e) which, as shown by recipient's records, was subsequent to disclosure, independently developed by an employee of recipient who neither directly nor indirectly had access to CONFIDENTIAL INFORMATION. Specific CONFIDENTIAL INFORMATION shall not be deemed to come under the foregoing exceptions merely because it is embraced by more general information which is or may become public knowledge.

This Agreement contains the entire understanding of the parties hereto, may not be changed except by another writing executed by all the parties, and shall be interpreted in accordance with the laws of the State of New Jersey. All amendments, changes or additions to this Agreement will be reduced to writing and will require the signature of both the SPONSOR and the INVESTIGATOR before becoming effective.

IN WITNESS WHEREOF, the parties have caused this Agreement to be duly executed in duplicate.

Z-Pharm, Inc.	Participant
Signature: _____	Signature: _____
Date: _____	Date: _____
Print Name:	Print Name:
Title:	Title:

Chapter 8

Study Protocol Preparation, Review, and Approval

Jeffrey Ambroso PhD, DABT*, Amy Babb BS†, Karen VanLare BA, RVT‡, James Greenhaw BS, LAT†, Kelly Davis DVM†, Joe M. Fowler BS, RQAP-GLP† and William F. Salminen PhD, DABT, PMP**

*Department of Safety Assessment, Glaxo Smith Kline, Research Triangle Park, NC, †National Center for Toxicological Research, FDA, Jefferson, AR, ‡Novartis Animal Health, Greensboro, NC, **PAREXEL International, Sarasota, FL*

Key Points

- The study protocol guides the technical conduct of the study and, therefore, should contain sufficient details about the study design and procedures
- The protocol should be thoroughly reviewed by the Contracting Scientist, Study Director, and other relevant personnel to ensure that all study details are captured and the protocol is free of logistical traps
- Changes to the protocol must be made through a protocol amendment and all study personnel should be made aware of the changes

---

This chapter deals with the preparation, review, and final approval of the Good Laboratory Practice (GLP)-compliant study protocol. The study protocol guides the technical conduct of the study and is designed to address a scientific question. The study protocol is a key component of compliance with GLPs as it provides a means to "reconstruct" the study and will document any changes or deviations to the original study plan. Therefore, it is essential that the protocol is clearly written and understandable by all study personnel and covers all of the key study functions that need to be conducted. Particular emphasis is placed on the logistical aspects of the study protocol and design flaws that set studies up for failure. The reader will be walked through the entire process starting from drafting the protocol to avoid logistical traps, communication and tracking of protocol changes, internal and external (e.g. regulatory authority) protocol review, and final approval. A generic protocol template is reviewed, which will help the reader design protocols that result in successful studies.

Nonclinical Study Contracting and Monitoring. http://dx.doi.org/10.1016/B978-0-12-397829-5.00008-9

WRITING THE FIRST DRAFT

One of the best ways to start the process of developing a draft protocol is to provide the laboratory with a detailed study outline (see Chapter 6). The study outline should cover all of the key study functions that need to be conducted during the study. It is often easiest if the laboratory writes the first draft using the details in the study outline since they have protocol templates and will be able to tailor the protocol to the laboratory's specific procedures, terminology, and capabilities. For example, a laboratory may use the term "clinical observations" to represent a limited set of observations that do not involve removing the animal from the cage; whereas, the term "detailed clinical observations" represents a thorough assessment of the animal after removing it from the cage. By having a detailed study outline that clearly lists the specific requirements, the laboratory can match their terminology to the study requirements. Once the laboratory has drafted the protocol, an easy way to review, make comments, and track changes is for the laboratory to provide the Contracting Scientist with an electronic version that can be marked up using track change tools (e.g. the track change function in Microsoft Word). This not only facilitates your review and commenting on the protocol, but allows the changes and comments to be clearly communicated back to the laboratory.

Even for the same study design, each laboratory is likely to have specific ways they conduct various study functions and the timing of those functions. This could be a result of their standard operating procedures, personnel limitations (e.g. limited manpower or experience), or laboratory limitations (e.g. limited space for conducting necropsies). If you have specific requirements that have to be done a certain way, you should clearly describe these in the study outline so that the laboratory understands that their "normal" procedures may have to be altered to meet the study requirements. For example, some laboratories feed canines once a day for a set period (e.g. 4 hours); whereas, others provide feed *ad libitum*. If you require *ad libitum* feeding, you must list this in the study outline so that the laboratory clearly understands this requirement. Another example is if you require all necropsies and terminal bloods to be conducted within a certain timeframe. If you do not specify the time window, the laboratory may conduct necropsies over the course of the entire day or multiple days leading to possible diurnal effects. In order to meet the time window, the laboratory may need to bring in additional personnel, set up additional necropsy stations, and/or have a staggered study start with staggered necropsies. The laboratory must write the protocol by combining the Sponsor's study requirements with the laboratory's procedures and capabilities. Unless the laboratory clearly understands the study requirements (i.e. by having a detailed study outline), it is unlikely that the draft protocol will meet all of the specific requirements.

A final note about the draft protocol is that the study outline should clearly state the regulatory, format, and contract requirements. For example, for a GLP study, the protocol should include all relevant GLP-required sections and places for approval (e.g. Study Director signature line). If the protocol will be submitted for regulatory review and approval, it may need to meet certain formatting requirements that differ from the laboratory's standard protocol template and/or include specific information (e.g. a title page containing certain information in a specified format). If the study contract requires certain approvals (e.g. Sponsor approval and signature), the protocol should include a place for clearly indicating the approvals.

REVIEWING THE DRAFT PROTOCOL

Once you receive the draft protocol, you must thoroughly review it to ensure that it meets your requirements. It is important to go through the protocol line by line so that you understand exactly what will be done during the study and ensure that it lines up with your requirements. Having an electronic copy of the protocol (e.g. Microsoft Word or Adobe PDF) allows you to make comments and/or changes easily. By using track change and commenting functions, the laboratory will be able to not only see your comments but also easily see your recommended changes.

If you are working with a laboratory for the first time, it is especially important that you not only review the protocol but ensure that you understand what the different terms mean and what will actually be done during the study. Reviewing the laboratory's Standard Operating Procedures (SOPs) for the procedures to be performed during the study is a good way to understand how a lab does things. As mentioned previously, the draft protocol may list that "clinical observations" will be conducted twice a day in the morning and afternoon at least four hours apart. You may assume that clinical observations involve removal of the animal from the cage and assessment of animal behavior and neurological function (e.g. ability to stand and walk normally). However, the laboratory's SOP may define clinical observations as essentially cage-side observations of the animal without removing it from the cage. Unless you fully understand what a given term means and what functions will be conducted (i.e. you have a copy of the laboratory's relevant SOP or know the laboratory's procedures), it is best specifically to ask the laboratory what they will do for a given study function.

One critical aspect of reviewing the protocol is ensuring that it is logistically feasible. Experienced laboratories will often warn you if certain study functions conflict with each other, are likely to be difficult to conduct, or may not meet the required time windows. However, some laboratories may not warn you of potential issues since they are afraid that you may not place the study with them or they may simply not have experience with the particular study design and not be able to predict logistical issues. This is where you should outline all of the study functions and timing requirements to ensure that there are no conflicts that are likely to lead to protocol deviations. For example, you may require hourly clinical observations for the first six hours after dosing along with pharmacokinetic blood draws at 15 and 30 minutes, and 1, 2, 3, 4, 5, and 6 hours. In addition, all dosing has to be completed for all animals within a one-hour window. Meeting all of these requirements is likely to be very difficult and you must ensure that the laboratory understands the requirements. Often some specifications are not absolutely required and some items may be altered in order to be able to conduct the study without incurring protocol deviations. If you think the timing of some functions may lead to study deviations, you should ask the laboratory if they can truly meet the timing requirements.

A useful method for outlining the study and identifying potential logistical conflicts is to draft a day-by-day study function chart (study calendar). This is the same chart provided with the detailed study outline (see the end of Chapter 6) but it is tailored to the specific laboratory and draft protocol. The study calendar lists all of the key study functions that need to be conducted on each day of the study. If study functions are conducted during the acclimation phase (e.g. physical examinations or baseline blood draws), the calendar should start during the acclimation phase but it can start as early as when animals arrive at the facility. The calendar can become very long for chronic studies, but it is a great way to determine if too many study functions will be conducted on certain days and are likely to conflict with each other.

The following is a general protocol template that can be used for most nonclinical studies. The template can be used to draft a GLP-compliant protocol; however, the main purpose is to serve as a reference for the Contracting Scientist about the main components of a protocol. There are many different types of nonclinical studies and some involve unique functions and assessments (e.g. reproduction and development protocols). This protocol template is not meant to cover every possible study design and function, but instead lay the foundation for the major components of a protocol. Each section of the protocol template is reviewed along with what should be included in the different sections and potential issues to be aware of. The sections that are required by the GLPs are specified; however, these are a bare minimum for a decent protocol and many additional sections are typically required to describe adequately how the study will be conducted.

Title Page

The following information is typically included in the protocol title page.

Study Title

The title should provide a clear description of the type of study, test system, and what is being tested. This is a GLP required section.

Testing Facility
Name and address should be provided.

Testing Facility Study Number

Sponsor Study Number
If the Sponsor has their own study number, this should be included for future cross-referencing.

Study Director
Name and title should be provided.

Sponsor
Company name and address should be provided.

Table of Contents
This helps to locate quickly various sections of the protocol. For example, a technician may want to know when physical examinations are being conducted and this can be quickly found via the table of contents without having to read through the entire protocol. A table of contents can be inserted and created easily in Microsoft Word.

Body of Protocol

Introduction

Study Title
Copy from title page.

Testing Facility
Name and address. Include any accreditations (e.g. AAALAC). This is a GLP required section.

Study Number

Testing Facility Study Number
Copy from title page.

Sponsor Study Number
Copy from title page.

Study Director

Name
Title
Address
Contact information (e.g. phone, cell, e-mail)

Sponsor

Copy from title page. This is a GLP required section.

Sponsor Representative(s)

Name
Title
Address
Contact information (e.g. phone, cell, e-mail)

Objective and Purpose

Provide a brief one to two sentence overview about the purpose of the study. This is a GLP required section.

Experimental Design Overview

Provide a high level one paragraph overview of the experimental design including treatments, key procedures, and observations.

Regulatory Compliance

Test Guideline

Indicate any relevant study guidelines that will be followed (e.g. OECD, ICH).

Good Laboratory Practice

Indicate if the study will follow GLPs and which ones (e.g. FDA, OECD, Japan).

Proposed Study Schedule

Study Initiation Date

This is the date the Study Director signs the protocol.

Experimental Start Date

See note below.

Experimental Termination Date

See note below.

Draft Report Mail Date

The experimental start and termination dates and draft report mail date can be indicated if known but often these dates are unknown when the protocol is signed/approved; therefore, they may have to be added by an amendment to the protocol.

Quality Assurance

Indicate Quality Assurance Unit's (QAU) functions during the study (e.g. periodic inspection of the study and review of the draft and final report according to their SOPs).

Alteration of Study Design

This section should clearly indicate that alterations to the protocol can be made as the study progresses but they have to be through a protocol amendment that is approved by the Sponsor. Amendments should be reviewed, approved, and signed in the same manner as the original protocol.

Submission of Study to Regulatory Authorities

Indicate if the study is expected to be submitted to a regulatory authority (e.g. to the US FDA to support a new drug application) and if it will be included on the laboratory's master schedule in accordance with the GLPs.

Test and Control Articles

Description of Test Article

This is a GLP required section.

Identity

List the test article name or other identifying information. If known, include lot number, storage conditions, expiration date, safe handling procedures, etc. or indicate that this information will be documented in the study data.

Test Article Properties/Characterization

Indicate who will provide documentation on the strength, purity, composition, stability, physical properties, and method of synthesis, fabrication and/or derivation of each batch of test article and that this information will be included in the study data. Since the Sponsor usually provides the test article, this information is typically provided by the Sponsor. It is important to note that the laboratory must have this information or a verification that the Sponsor has this information otherwise it is a GLP deviation. Even if the Sponsor verifies that they have this information, the laboratory may still list this as a GLP deviation in the compliance statement unless they have the information to include in the study file. Also, a test article manufactured under Good Manufacturing Practices (GMP) technically does not comply with the GLP requirements and a laboratory may list this as a GLP deviation in the compliance statement even though the test article was manufactured under tightly controlled conditions and was fully characterized.

Test Article/Dose Formulation Preparation

Provide information on how the test article will be prepared for dosing (e.g. administered as received, mixed with a vehicle, mixed in feed) along with any special preparation, mixing, or handling instructions.

Test Article/Dose Formulation Analysis

Indicate what analyses will be conducted to ensure the test article and/or dose formulations are stable, homogeneous, and meet the required concentrations throughout the course of the study.

Reserve Sample

Indicate what reserve samples (e.g. a sample from each batch) will be collected and archived.

Test Article Disposition

Indicate the fate of the remaining test article at the end of the study (e.g. returned to Sponsor or destroyed).

Description of Control Article

This is a GLP required section.

Identity

List the control article or other identifying information. If known, include lot number, storage conditions, expiration date, safe handling procedures, etc. or indicate that this information will be documented in the study data.

Control Article Properties/Characterization

Indicate who will provide documentation on the strength, purity, composition, stability, physical properties, and method of synthesis, fabrication and/or derivation of each batch of control article and that this will be included in the study data.

Safety Precautions

Indicate what safety precautions should be taken for working with the test and control articles (e.g. specific personal protective equipment).

Test System

This is a GLP required section.

Species
Strain or Breed

Source

Indicate where the animals will come from (e.g. name of a commercial breeder, in-house breeding facility).

Justification of Test System

Provide a rationale why alternatives are not applicable (e.g. *in vitro* systems) and why the specific species and strain were selected. Normally this section will refer to regulatory requirements to conduct the study in an *in vivo* system.

Expected Age

Indicate the age range of the animals. If the animals have to be a certain age on the first day of dosing (e.g. ≤6 weeks of age), this needs to be clearly indicated.

Expected Body Weight

Indicate the body weight range of the animals. If the animals have to be a certain weight on the first day of dosing (e.g. 300–400 g), this needs to be clearly indicated. Typically, the body weight is driven by the age of the animals and is simply documented in the study data. Note that if the weight range is too narrow, you may have protocol deviations or have to exclude animals at the extremes. It is best to use a large weight range to avoid this.

Number of Animals on Study

Number Ordered Often, this number is greater than the number enrolled since extra animals are often ordered to ensure that only animals meeting the inclusion criteria are enrolled.

Males
Females
Number Enrolled on Study

Males
Females
Justification for Number of Animals on Study

A short explanation should be provided for why the number of animals is needed for the study. For example, the study was designed to use the fewest number of animals possible and meet the objectives of the study, scientific needs of the Sponsor, required statistical power, and regulatory requirements.

Inclusion/Exclusion Criteria

Clear inclusion and exclusion criteria for enrolling the animals on study should be provided (e.g. only healthy animals meeting the specified age and weight ranges will be enrolled).

Assignment to Groups

A description should be provided as to how animals will be assigned to groups in a manner that controls bias (e.g. using a randomization procedure). The fate of the extra animals not officially enrolled in the study should also be indicated (e.g. euthanized, held until completion of the study).

Method of Identification

The method of identification should be listed (e.g. ear or tail tattoo, ear tag, cage label, radio frequency identification [RFID] implant).

Final Disposition

The final disposition of the animals should be indicated (e.g. remaining carcass incinerated after necropsy), including the fate of the extra animals not enrolled in the study (e.g. euthanized and incinerated, returned to general animal colony).

Husbandry

Acclimation

Information on the duration of the acclimation period and any study functions that will be conducted (e.g. physical examinations, body weight, baseline blood draws should be listed).

Housing and Environmental Conditions

The type of housing (e.g. polycarbonate cages with hardwood bedding, runs with raised flooring) should be specified along with the environmental conditions (e.g. temperature and humidity ranges, light/dark cycle, number of animal room air changes per hour).

Environmental Enrichment

The type of environmental enrichment should be specified. For large animals that require exercise periods under the animal welfare regulations (e.g. canines), the method for meeting this requirement should be listed. If environmental enrichment will not be used, the rationale should be listed.

Diet

Indicate the diet that will be fed to the animals and how often the feed will be provided (e.g. *ad libitum*, 4 hours per day). Fasting periods, if used, should be listed in this section with specifications about the duration (e.g. overnight for

8–12 hours). Also, indicate that each lot utilized will be identified in the study records along with any relevant analyses (e.g. certificate of analysis for certified feed). This is a GLP-required section.

Contaminants

Indicate if there are any contaminants in the feed that may interfere with the study. Typically for certified feed, the analysis provided in the certificate of analysis (e.g. pesticides, heavy metals) is sufficient.

Water

Indicate the source of the water (e.g. municipal water supply, on-site well) and how often it will be provided to the animals (e.g. *ad libitum*).

Contaminants

Indicate if there are any contaminants in the water that may interfere with the study. Also indicate how often drinking water will be monitored for specific contaminants (e.g. for contaminants and at intervals as specified in the laboratory's SOPs). The results of the relevant analyses should be maintained in the study or facility data.

Experimental Design

A complete description of the study design including the number of groups, doses, number of animals, etc. should be provided. Including a table similar to Table 8.1 that outlines the study design is useful.

Information should be provided on the methods that will be used to control bias (e.g. animals' weight ranked and then randomly assigned to groups). This is a GLP required section.

Test and Control Article Administration

This is a GLP required section.

Route of Administration

Indicate how the test article will be administered (e.g. oral gavage, topical, IV injection).

Frequency of Dosing

Indicate how often the test article will be administered (e.g. once per day).

Duration of Dosing

Indicate how long the test article will be administered (e.g. 28 days). If there are satellite recovery groups or dosing holidays (e.g. weekends or defined developmental periods), these should be clearly indicated.

TABLE 8.1 Study Design Outline

Group	Number of animals	Test article	Dose formulation concentration (mg/ml)[1]	Dose volume (ml/kg)	Dose (mg/kg)[1]	Duration	Fasting[2]	Necropsy
1	5 m / 5 f	0.5% MC	NA	10	NA	1X/d for 28d	Overnight prior to each dose	24h post 28th (last) dose
2	5 m / 5 f	ABC-123 in 0.5% MC	1	10	10	1X/d for 28d	Overnight prior to each dose	24h post 28th (last) dose
3	5 m / 5 f	ABC-123 in 0.5% MC	3	10	30	1X/d for 28d	Overnight prior to each dose	24h post 28th (last) dose
4	5 m / 5 f	ABC-123 in 0.5% MC	10	10	100	1X/d for 28d	Overnight prior to each dose	24h post 28th (last) dose

[1] It should be clearly indicated how the dose formulation concentrations and doses are expressed. For example, the entire compound (e.g. active + salt + water) or just the active ingredient. If the values are expressed as active ingredient, the conversion factor used to extrapolate the whole compound values to active ingredient should be listed.
[2] Animals will be fasted starting at 5 p.m. each night. Animals will also be fasted prior to necropsy.
MC = methylcellulose

Dose Levels

The dose levels should be listed again even if they were provided in the Experimental Design section/chart (e.g. 0, 10, 30, and 100 mg/kg). It should be clearly indicated how the dose formulation concentrations and doses are expressed. For example, the entire compound (e.g. active + salt + water) or just the active ingredient. If the values are expressed as active ingredient, the conversion factor used to extrapolate the whole compound values to active ingredient should be listed.

Justification of Route of Administration, Frequency and Duration of Dosing, and Dose Levels

Provide rationales for the route of administration (e.g. the test article is an investigational drug proposed for oral use in humans), the frequency and duration of dosing (e.g. needed to meet regulatory requirements), and dose levels. If previous studies exist in the same species (or even other species), they can be used to support the dose level selection.

Procedure

Describe how individual animal doses will be determined and conducted. For example, the dose volume will be based on the daily body weight taken prior to dosing, doses will be rounded to the nearest graduation on the syringe, the daily dose will be given by oral gavage in the a.m., and the first day of dosing will be designated as Study Day 1.

In-Life Study Evaluations

This section of the protocol should list all of the relevant study functions that will be conducted while the animals are alive. This list is not all inclusive and is meant to cover most general nonclinical toxicology study designs. Specialized studies (e.g. reproduction and developmental studies) will have many unique evaluations that are not listed here. The full detail of how these procedures are done can be in an SOP, so this section does not necessarily describe these in detail. This is a GLP required section.

Physical Examinations

Indicate when physical examinations will be conducted and by whom (e.g. a staff veterinarian). Also indicate what the examination will entail (e.g. general condition of the animal with a visual check of body condition, hair coat, attitude, ambulation, eyes, and oronasal regions).

Ophthalmology Examinations

Indicate when eye examinations will be conducted and by whom (e.g. a board certified veterinary ophthalmologist).

Cageside Observations

As mentioned previously, different laboratories have different terminologies for these types of observations. Here, they are defined as looking at the animal through the cage to assess morbidity, mortality, injury, and availability of food and water. Indicate how often these will be conducted (typically at least twice a day) and what the course of action is for animals in poor health (e.g. notification of the veterinarian and Study Director along with additional monitoring).

Detailed Clinical Observations

As mentioned previously, different laboratories have different terminologies for these types of observations. Here, they are defined as removing the animal from the cage and conducting a thorough assessment of the animal's health including behavior. The laboratory should have an SOP that clearly defines what assessments will be conducted as part of the detailed clinical observations and you should ensure that they line up with your requirements. Indicate how often these will be conducted and what the course of action is for animals exhibiting abnormal signs (e.g. notification of the veterinarian and Study Director along with additional monitoring).

Unscheduled Deaths and Moribund Animals

It is often useful to include information on what the course of action is for animals that are moribund (i.e. near death) or that are found dead. For example, moribund animals can be sent to be examined immediately by a veterinarian, euthanized, and then a necropsy with tissues collected according to the protocol or they can be euthanized and discarded. Dead animals often have extensive organ autolysis and little information can be derived from a histopathological assessment of the organs; however, some studies may require that even found dead animals are sent to necropsy and organs preserved for analysis.

Body Weight

Indicate how often body weight will be measured.

Feed Consumption

Indicate how often feed consumption will be measured.

Water Consumption

Indicate how often water consumption will be measured.

Clinical Pathology

Indicate if animals will be fasted prior to blood collection, approximately how much blood will be collected, the source of the blood (e.g. peripheral

vein), and if the withdrawal amounts for all blood draws are within acceptable limits (consult Diehl et al., 2001[1] for typical limits). Also indicate if bleeding needs to follow a specified order to avoid temporal bias
(e.g. one animal per group and then cycling back to the first group) or
collected within a specified time window (e.g. within a 3-hour period between
8 and 11 a.m.).

Hematology

Number of Animals

Indicate which animals will be bled (e.g. all animals).

Collection Intervals

Indicate at what timepoints blood will be collected (e.g. pretest and the day of
necropsy).

Sample Type

Indicate the type of collection tube that will be used (e.g. EDTA tube).

Parameters Evaluated

List the specific tests that will be conducted. If sample volume may be limited,
you should indicate a priority order for the sequence of tests so that the highest
priority tests are run first until the sample runs out.

Clinical Chemistry

Number of Animals

Indicate which animals will be bled (e.g. all animals).

Collection Intervals

Indicate at what timepoints blood will be collected (e.g. pretest and the day of
necropsy).

Sample Type

Indicate the type of collection tube that will be used (e.g. serum separator tube).

Parameters Evaluated

List the specific tests that will be conducted.

---

1. Diehl, K.H., Hull, R., Morton, D., Pfister, R., Rabemampianina, Y., Smith, D., Vidal, J.M., and
vandeVorstenbosch, C. (2001). A good practice guide to the administration of substances and
removal of blood, including routes and volumes. *Journal of Applied Toxicology* 21: 15 – 23.

Urinalysis

Indicate how urine will be collected (e.g. animals in metabolism cages, pans under the cage, cystocentesis) along with the duration of urine collection (e.g. overnight).

Number of Animals

Indicate which animals will have urine collected (e.g. all animals).

Collection Intervals

Indicate at what timepoints (e.g. pretest and the day of necropsy) and how (e.g. cystocentesis, cage pan, metabolism cage) urine will be collected.

Parameters Evaluated

List the specific tests that will be conducted.

Scheduled Euthanasia

Indicate the method of euthanasia and how death will be verified. It is also useful to list if the method of euthanasia is conducted in accordance with the American Veterinary Medical Association (AVMA) Guidelines on Euthanasia.

Postmortem Evaluations

Indicate the type of necropsy that will be performed (e.g. complete necropsy), who will supervise the necropsy (e.g. a board certified veterinary pathologist), and which animals will be subject to the procedure (e.g. all animals [including moribund and found dead animals], only animals surviving to scheduled necropsy). This is a GLP required section when necropsies will be part of the study.

Gross Necropsy

Indicate the procedures and examinations that will be conducted as part of the gross necropsy and who will conduct the gross necropsy and organ collections (e.g. qualified Testing Facility personnel). Indicate the fixatives that will be used to preserve various organs (e.g. some organs such as the eyes and testes are better preserved in special fixatives such as Modified Davidson's fixative instead of neutral buffered formalin). It is also best to indicate clearly if certain organs are handled a certain way to ensure adequate fixation (e.g. lungs and bladder inflated with fixative), identification (e.g. distinguishing paired organs), and/or adequate observation (e.g. opening and observing the entire gastrointestinal tract and whether this would be done at necropsy or during the trimming phase of tissue preparation). This section should also discuss what will happen with gross lesions that are observed (e.g. recorded and preserved for histopathological analysis).

Organ Weights

Indicate what organs will be weighed, how paired organs will be handled (e.g. weighed together), and if organs will be weighed fresh at necropsy or post-fixation. Indicate if additional weight calculations will be made (e.g. organ relative to body and brain weights). Indicate if organ weights will be collected for moribund and/or found dead animals. Organ weights are typically not collected from moribund or found dead animals since there will be no control to compare the values to and as the animals are growing there will be differences based on age alone.

Histopathology

It is useful to use a table similar to Table 8.2 that lists the specific organs/samples that will be collected. Also, indicate what groups will be read (e.g. only tissues from the control and high dose groups will be read unless a target organ(s) is identified and then the target organ from the lower dose groups will be processed and read).

Indicate how the fixed tissues will be processed (e.g. hematoxylin and eosin stained paraffin sections) and read (e.g. microscopic examination by a board

TABLE 8.2 List of Organs that will be Weighed, Preserved, and Examined

Tissue	Organ weight taken	Collected and preserved	Microscopic examination (All groups)
Adrenal gland	X	X	X
Aorta		X	X
Bone with bone marrow, femur		X	X
...list of organs continues...		X	X
Heart	X	X	X
...list of organs continues...		X	X
Liver	X	X	X
...list of organs continues...		X	X
Vagina		X	X
Gross lesions/tissue masses		X	X

certified veterinary pathologist). Indicate if the pathologist can use special stains to help with the diagnosis of findings and if Sponsor approval is required prior to the use of additional stains.

Statistical Methods

Provide information on the types of statistical analyses that will be conducted on the different types of data (e.g. means and standard deviations, simple descriptive statistics, analysis of variance [ANOVA]). Also, it is important to indicate the statistical unit, especially for reproduction and developmental studies (e.g. individual animal, litter). This is a GLP required section.

Study Reports

Progress Reports

Indicate the frequency of any progress reports that will be sent to the Sponsor (e.g. monthly throughout the study) and how they will be provided (e.g. electronic file via e-mail, secure portal, quality-checked).

Final Report

Indicate the general contents of the draft report, the format (e.g. electronic, paper), if it will be QAU audited prior to Sponsor review, and when it will be submitted to the Sponsor (e.g. 4 months after the completion of the necropsy). Indicate that a final report will be issued once all comments from the Sponsor have been adequately addressed. Specify the format of the final report (e.g. electronic and/or paper) and number of copies.

Records and Specimen Retention

Provide information on what records, data, and samples will be retained and the duration of the retention. Most laboratories will store paper data and samples for a short period and then charge a fee for continued storage. This is a GLP required section.

Compliance with Animal Welfare Regulations

This section should provide information on how the study complies with applicable mandatory and voluntary animal welfare regulations (e.g. USDA AWA, AAALAC, The Guide). This section should also list what assessments were made to ensure that the study is needed (e.g. required for a regulatory submission), does not duplicate previous experiments, ensures the highest level of animal welfare, and cannot be conducted using alternative means (e.g. *in vitro* methods). It is also useful to state clearly that the protocol and any amendments or procedures involving the care or use of animals will be reviewed and approved by the Testing Facility IACUC and possibly the Sponsor IACUC prior to the initiation of the procedures and that any animals that are in overt pain/distress or appear moribund and are unlikely to recover will be humanely euthanized.

Signatures

Date of Sponsor Approval

The date of Sponsor approval is required under GLPs. Sponsor signature is not required under GLPs but it is a good way to show that the Sponsor reviewed and approved the protocol.

Dated Signature of Study Director

The Study Initiation date is the date the Study Director signs the protocol. This is a GLP required section.

Test Facility Management

Some laboratories may not have Test Facility Management sign the protocol, but this is a good way to show that management reviewed and approved the protocol. This can help show that Test Facility Management is adequately assessing Study Director Workload and study resources, both of which are required under GLPs.

After you have reviewed the draft protocol, you will want the Study Director to make changes and address your comments before submitting it for further review. Once you and the Study Director are satisfied with the protocol, there are additional reviews that may be needed: internal to your company, at the laboratory, and/or at a regulatory authority. Your company may have certain reviews that must be completed before the protocol is finalized. Some companies have their own QAU that will review the protocol for GLP compliance and some companies even have animal welfare officers that will review the protocol to ensure compliance with regulatory-driven and company-driven animal welfare requirements. In addition, other scientists in your company may need to peer-review your protocol. If your study needs to start by a specified date, it is important that you take into consideration these additional reviews since not only will they take a while to complete but comments and suggested changes arising from the reviews may take a while to resolve, especially if they conflict with each other (e.g. animal welfare requirements conflict with the scientific needs of the study).

At the laboratory, the Study Director should have already solicited input on various aspects of the protocol if he/she was unsure how certain functions would be conducted and coordinated. However, once the protocol is in a final draft form, it should be circulated to all relevant functions (e.g. in-life, formulations, clinical pathology, pathology) not only to ensure that each function is adequately covered but also that the coordination of functions flows smoothly. For example, the in-life group should coordinate with the pathology group about when animals need to arrive at the pathology facility since various in-life functions may have to be completed (e.g. body weight, clinical observations, removal from the in-life computer system) before animals are removed from the animal room.

A critical review that needs to be completed at the laboratory is the IACUC review. This review will ensure that the study complies with all relevant animal welfare requirements. Each facility IACUC has different methods of reviewing the protocol and there is no one right way as long as the method complies with the applicable animal welfare regulations. For example, some facilities will have every IACUC member review every protocol; whereas, others will have only several designated IACUC members review a given protocol. If concerns arise, then the designated reviewers can have the entire IACUC review the given protocol. The important thing to remember is that the IACUC review ensures that the study is scientifically justified from an animal use perspective and complies with animal welfare requirements. If the protocol is discussed during an IACUC meeting, it is often helpful for the Contracting Scientist to participate in the discussion (e.g. via teleconference or in person). This allows the Contracting Scientist to answer any questions that may come up that the Study Director may not be able to answer (e.g. does the study duplicate previous work?; what is the rationale for the high dose selection?; why can't *in vitro* studies be used?). Since the IACUC meeting is a closed meeting, the Contracting Scientist will only be able to participate in the portion discussing their protocol and will be excused so that the IACUC can discuss the protocol in private. Another thing to keep in mind is that IACUCs meet at different frequencies. If an IACUC only meets monthly and your protocol is designated for full IACUC review, finalization of your protocol may be delayed until the next IACUC meeting. Fortunately, most commercial laboratories understand the importance of timely IACUC review and have frequent meetings and/or procedures to expedite protocol reviews. Once the IACUC review is complete, you may have to revise the protocol to address any questions or concerns followed by resubmission until final IACUC approval is received.

Depending on the type of protocol and purpose of the study, the protocol may need to be reviewed by a regulatory authority before the study can start. In these situations, you can always run the study without regulatory authority sign off but you run the risk of having a study design that may not be accepted by the regulatory authority once you submit it to support a product approval. When a protocol needs regulatory authority review, it should have gone through all of the previously-mentioned company and laboratory reviews and be in a final draft form that does not require any changes from the Sponsor's or laboratory's perspectives. If the protocol is found to be acceptable by the regulatory authority, then no changes should be made to the protocol otherwise the changes may make the study unacceptable from the regulatory authority's perspective.

If changes to the protocol are required after regulatory authority approval, they should be reviewed and approved by the regulatory authority. Also, if the regulatory authority requests specific changes to the protocol, the changes should be made or explanations provided as to why certain changes cannot be made and the protocol submitted for review again. Once the regulatory authority provides a final sign off, then the protocol can be finalized.

FINALIZING THE PROTOCOL

Developing a final nonclinical study protocol can range from relatively simple and quick to very complex and time consuming depending on the study design and company, laboratory, and regulatory authority review procedures. The key to developing a solid protocol that meets the scientific and animal welfare requirements and is free of logistical traps is ensuring that all relevant groups have had sufficient time to review the protocol. This ensures that the study functions can be conducted correctly without incurring protocol deviations and that coordination between different groups has been thoroughly worked out. Rushing the development of a protocol is a recipe for disaster since it does not allow all of the study details to be thoroughly worked out. At best, a rushed protocol is likely to have many protocol amendments to adjust for unanticipated issues. However, it is likely that a rushed protocol will have many study deviations due to poor planning and these could ultimately compromise the integrity of the study. Therefore, it is critical to take sufficient time to develop a protocol and have it reviewed and re-reviewed by the relevant laboratory functions to ensure the study will run smoothly.

Once the protocol is in its final form, it needs to be approved. Under GLPs, the Study Director approves the study protocol and initiates the study by signing and dating it. Although Test Facility Management does not have to review and approve the protocol under GLPs, it is a good practice so that they can adequately assess Study Director workload and ensure resources are available to conduct the study, both of which are GLP requirements. Signature by Test Facility Management is an easy way to show clearly that management reviewed and approved the protocol. The protocol must list the date of Sponsor approval, which can be via a variety of means such as e-mail, fax, or other documented means. Most laboratories will have the Sponsor (Contracting Scientist) sign and date the protocol as part of the approval process even though a signature is not technically required under GLPs. Once finalized, a copy of the signed protocol should be sent to the Contracting Scientist. At the laboratory, the final protocol should be readily accessible by all applicable study personnel. In addition, the laboratory should use a method to distinguish draft from final protocols to ensure that only the finalized protocol is used to conduct the study. One way to do this is to print finalized protocols on a uniquely colored paper. If protocols are available electronically, a method must be used to ensure that only the final approved protocol is used by study personnel (e.g. draft versions are retained on a computer or network drive with limited access [e.g. Study Director and Test Facility Management only] and only the final signed version, in an uneditable format [e.g. Adobe PDF], is available to study personnel).

CHANGING THE FINALIZED PROTOCOL

Many studies require changes to the protocol after the protocol is finalized. This can occur well before any study functions are conducted or as the study progresses. For example, the protocol may be finalized without knowing the exact dates that animals will be enrolled, dosed, and sent to necropsy. Once these dates are known, they are added to the protocol via a protocol amendment. Once dosing is underway, you can run into unexpected situations where study functions have to be removed, changed, and/or added. For example, the protocol may specify clinical observations be conducted twice a day in the morning and afternoon at least four hours apart. After the first dose, you notice that the animals are experiencing adverse effects and you want to conduct more frequent hourly observations after each dose. These would be added to the protocol by a protocol amendment. You may also run into a situation where unexpected logistical issues are causing a large number of protocol deviations. Protocol deviations will be discussed more in a subsequent chapter but they are essentially study functions that did not meet the protocol- or SOP-driven specifications. For example, pharmacokinetic blood draws had to be taken at 5, 15, 30, 60, and 90 minutes from a large number of animals along with the conduct of other study functions. Due to difficulty in accessing a vein, the time windows were not met on the first day of blood draws resulting in a large number of protocol deviations. A protocol amendment was issued to drop the 5-minute blood draw and also use a temporarily placed catheter instead of the previously protocol-specified individual peripheral vein sticks.

The protocol amendment should clearly state what is being revised and the justification for the revision. The protocol amendment should be reviewed and signed by all of the same parties who signed the protocol. In addition, amendments involving animal care and use procedures must be approved by the IACUC prior to performing the new procedures or functions. The protocol amendment should be issued in the same manner as the protocol and a copy should be placed with every copy of the protocol. If a protocol amendment is not readily available with the protocol, then the study personnel will not know what has changed with the study leading to protocol deviations. It is important to ensure that the laboratory has a robust method of ensuring that study personnel are aware of all protocol amendments, particularly ones that have a direct affect on how study functions are conducted. It is ultimately the Study Director's responsibility to ensure that study personnel know how to run the study, but having appropriate procedures in place to ensure study personnel are up to date on the protocol and amendments is a good way to avoid protocol deviations.

PROTOCOL CHECKLIST

PROTOCOL CHECKLIST	YES	NO	N/A or Comment
Drafting and Reviewing the Protocol			
1. A detailed study outline or other similar information was provided to the laboratory to facilitate the drafting of the protocol			
2. Protocol contains at a minimum all of the following GLP required sections: Title: Purpose Test and Control Article Identification Test Facility Name and Address Sponsor Test System: (include number, body weight range, sex, source, species, strain, substrain, age) Test System Identification Experimental Design: (include methods for the control of bias) Diet: (include information on acceptable levels of contaminants and mixtures with test article) Dosing: (dosage levels and frequency) Assessments: (type and frequency of tests, analyses, and measurements) Statistical Methods Records Date of Approval by the Sponsor Dated Signature of Study Director			
3. Draft protocol has been thoroughly reviewed by all relevant laboratory functions to ensure the smooth conduct and coordination of the study			
4. Protocol has been thoroughly reviewed by the Contracting Scientist to ensure that it meets the scientific/regulatory requirements			
5. Draft protocol has been reviewed by relevant company-specific functions (e.g. QAU, Animal Welfare, Scientific Peer-Review)			
6. Changes and comments are made in a manner that can be easily interpreted by the laboratory (e.g. electronic track changes)			
Finalizing the Protocol			
1. Does the protocol require regulatory review? If so, is the protocol in its final draft form (all relevant reviews have been completed and no additional changes are required)?			
2. Sponsor (Contracting Scientist) has reviewed and approved final protocol and the date of approval is included in the final protocol			
3. Test Facility Management has reviewed and approved the final protocol. How is approval documented?			
4. Study Director signed and dated the final approved protocol			
5. Only the final signed protocol is distributed to and accessible by laboratory personnel. What methods are used to ensure only the approved protocol is used to conduct the study?			
6. A copy of the final signed protocol is sent to the Sponsor			

PROTOCOL CHECKLIST	YES	NO	N/A or Comment
Protocol Amendments			
1. All changes to the protocol are made through a protocol amendment *prior to* implementation of the change			
2. Protocol amendments are reviewed, approved, and signed in the same manner as the approved protocol			
3. *All* protocol amendments are readily available with each copy of the approved protocol			
4. How are study personnel made aware of changes arising from protocol amendments?			
5. Copies of all protocol amendments are sent to the Sponsor			

Chapter 9

Test Article

Amy Babb BS*, William F. Salminen PhD, DABT, PMP†
and Joe M. Fowler BS, RQAP-GLP*

*National Center for Toxicological Research, FDA, Jefferson, AR, †PAREXEL International, Sarasota, FL

> ## Key Points
> - Test and control article sourcing is critical to the success of the study and all articles need to be fully characterized under GLPs
> - Test article should be synthesized using procedures similar to the final synthesis procedure, whenever possible, in order to maintain similar chemical profiles (e.g. impurity levels, physical form)
> - Mixtures of test article with a vehicle need to be characterized for stability, concentration, and homogeneity

This chapter covers the critical aspects of obtaining the test material for the study. Obtaining test material for a study seems relatively simple; however, there is critical documentation that needs to be obtained and analyses conducted to ensure Good Laboratory Practice (GLP) compliance. If the test material is mixed with feed or a vehicle, then additional testing of the mixtures is needed to ensure that they meet the required concentrations and are stable and homogeneous. All of the testing that is conducted is used to confirm that animals actually receive the stated doses and not lower or higher doses due to poor stability, contaminants, or non-homogeneous mixtures.

TEST AND CONTROL ARTICLE SYNTHESIS AND SOURCING

Many nonclinical studies use test article that is sourced directly from the Contracting Scientist's company. Each company has different procedures for sourcing test articles and you should follow those procedures to ensure that you receive the amount of test article you require in a timely manner. Depending on the state of product development, the test article may be readily available or it may have to be synthesized. The test article should be synthesized using procedures that will be as similar to the expected final synthesis procedure as

Nonclinical Study Contracting and Monitoring. http://dx.doi.org/10.1016/B978-0-12-397829-5.00009-0
2013, Published by Elsevier Inc.

possible. This helps ensure that the level and types of impurities and contaminants are similar to the product that will be ultimately sold. If a different synthesis procedure is used to produce the test article for the nonclinical studies, it is likely that the level and types of contaminants and impurities will be different from the final marketed product. In some situations, even if the contaminant and impurity levels and types are the same or the product is very pure (e.g. >99.5%), the physical form of the test article may be different (e.g. different polymorphs or micronized). The problem with these situations is that the animals in the nonclinical studies would be exposed to a different type of test article than the final marketed product possibly rendering the human risk assessments based on the nonclinical studies invalid (e.g. the final marketed product may have an additional impurity present that was not in the test material used for the nonclinical studies and the toxicity of the impurity would be unknown). Therefore, it is best to use a synthesis procedure for the test article that is as close to the final synthesis procedure as possible. Since some nonclinical studies are conducted at a very early stage of product development, well before the final synthesis procedure is known, a best guess should be made, the synthesis procedure thoroughly documented, and the test article fully characterized.

Determining the amount of test article to order or have synthesized can be difficult, especially if many studies need to be conducted and the doses are unknown. In this situation, you will have to estimate what the highest likely dose will be for each study along with the lower dose group levels. It is best to err on the side of selecting doses that are high since this will ensure that you do not run out of test article during the studies. Once you have estimated the doses, you need to determine the different studies that you need to run. For each study, you need to outline the number of dose groups, number of animals, expected weight of the animals (if dosing on a body weight basis), frequency of dosing (e.g. once daily, twice daily), and duration of dosing (e.g. 28 consecutive days). As with the dose estimation, it is best to estimate animal body weights on the high side to ensure you do not run out of test article. With this information, you can calculate how much test article you need for each study and the total amount of test article for the entire package of studies. It is always best to overestimate the final amount by increasing the amount by 10–25% since there is typically wastage if dose formulation mixtures need to be prepared; additional mixture volume is typically prepared to ensure an adequate amount for each dosing (e.g. if 10 ml of a dose mixture is needed to complete a daily dose for all animals, it is likely that the formulations department will prepare 12–15 ml of the mixture). Once you have your final test article amount requirement, you can work through your company sourcing procedure to obtain the test article. If the test article is difficult to synthesize or expensive, you may only be able to obtain the exact amount of test article you request. You may even be asked to revise your estimates without incorporating overestimates due to extreme limited availability of the test article. However, if test article can be synthesized in a large batch that is much greater than you need for all of your studies (e.g. you need 1 kg but the synthesis group can easily prepare 10 kg), it is often best to have the large batch synthesized not only to ensure you have sufficient test article for your planned studies but also for any future studies that may be needed. This makes comparison of results obtained from different studies much easier since the same batch of test article will have been used for all of the studies.

If you are having the control article sourced from your company, all of the above information about the test article also applies to the control article. However, typically, control article is purchased from a commercial supplier (e.g. methylcellulose as a suspension aid for oral gavage dosing is often purchased from a chemical supplier such as Sigma-Aldrich). Test article may also be purchased from a commercial supplier. In both situations, it is best if you can obtain sufficient test and control article from a single batch instead of multiple batches. For all batches that are purchased, you should obtain certificates of analyses (CofA) that provide sufficient characterization of the articles (e.g. identity, strength, purity, contaminants, expiration date).

Under GLPs, each storage container must be appropriately labeled and stored under conditions that maintain test article integrity. This is important not only during storage at the laboratory but also during shipment from the supplier to the laboratory. One thing to keep in mind when shipping test article is potential effects of the environment on the integrity of the article. For example, if test article is shipped during the summer in an uncontrolled environment (e.g. regular delivery truck), the temperature of the test article may reach a level that causes degradation. Therefore, means should be used to maintain the required environment or ensure that no alteration of the test article occurred (e.g. stability verification after the product is received at the laboratory, min/max thermometers in the shipping container). Another GLP requirement is that storage containers must be assigned to a given test article for the duration of a study.

A final note about sourcing test article is that reserve samples of the test and control articles must be retained for studies that are more than four weeks long. A sufficient amount of article should be retained for future analytical characterization of the article, if needed. Often this amount is relatively small compared to the amount needed for the study, but this still should be taken into consideration when determining how much article to order. If feed is used as a control article, then reserves of the feed need to be retained.

TEST AND CONTROL ARTICLE CHARACTERIZATION

Under the GLPs, the identity, strength, purity, and composition of the test and control articles must be determined and documented for each batch. The stability of the test and control articles must also be determined. In addition, methods of manufacturing the articles must be documented. As mentioned in Chapter 2 on GLPs, this can be a contentious issue with some laboratories since they require this information, or a statement from the Sponsor stating that this information is on file at the Sponsor, before initiating a study or they will list it as a GLP deviation. Regardless, this information must be available for each batch of test or control article that is used in the study. Therefore, it is best to obtain this information when you obtain the test or control article. For test articles, a CofA will meet the first two sets of requirements so long as it clearly lists the identity, strength, purity, composition, and stability of the article and the work was conducted under GLPs. It is important to obtain a CofA for each new batch of test article that you use for a study (e.g. if multiple batches are used for a study, then multiple CofAs will have to been obtained and kept on file). In order to comply with the GLP requirement, it is easiest if a copy of each relevant CofA is kept in the study records. If a CofA is not available for the test article, then analytical verification of the identity, strength, purity, composition, and stability of the test article is required and the work must be conducted under GLPs. As with a CofA, copies of this information should be retained in the study records for easy access and review by a regulatory authority if the study data are selected for audit during an inspection. Test article manufactured under Good Manufacturing Practices (GMP) technically does not comply with GLP requirements and a laboratory may list this as a GLP deviation in the compliance statement even though the test article was manufactured under tightly controlled conditions and was fully characterized. Stability of the test article does not have to be known prior to starting the study; however, in those cases, it must be determined during the study. For control articles that are currently marketed (e.g. methylcellulose sold by Sigma-Aldrich or certified feed used as a carrier), product labeling meets the requirements; however, it is best if you also obtain a CofA for each batch of control article. Copies of the product labeling and CofA should be retained in the study records. Since the laboratory often purchases the control article, it is important to ensure that they retain the required information in the study records.

The methods of manufacturing the test and control article must also be documented. If the control article is a currently-marketed product, then the product labeling and/or CofA meet this requirement. For the test article, it is best if you obtain documentation that clearly shows how the test article was manufactured. Unlike the analytical work characterizing the test article, the actual manufacturing does not have to be done or documented under GLPs. The Sponsor often retains the manufacturing information; however, it may be easiest to provide the information to the laboratory for inclusion with the study records so that it can be easily accessed during a regulatory review of the study. If the Sponsor retains the information, it may be difficult tracking down the details in a timely manner if a regulatory reviewer requires the information.

MIXTURES OF TEST ARTICLES

Many studies use a mixture of the test article with a vehicle to facilitate dosing. For example, some poorly soluble test articles are suspended in an aqueous cellulose derivative solution (e.g. 0.5% w/v methylcellulose) to maintain a homogeneous suspension during oral dosing. If a simple aqueous vehicle was used (e.g. water), the test article may separate too quickly leading to a non-homogeneous suspension. For studies administering the test article via the feed, the test article is mixed directly into the feed. Whenever a mixture is prepared for dosing, the uniformity, concentration, and stability of the mixture needs to be determined periodically under GLPs. The actual frequency is left up to the laboratory/Sponsor but should be sufficient to assess adequately the various properties of the mixture. Unlike test article characterization, each batch of a mixture does not have to be tested, just a sufficient number to verify the various required properties.

Whenever a mixture is selected for testing, a sample is collected and sent to the analytical laboratory for verification of the mixture's properties. The analytical work must be conducted under GLPs. A GLP analytical method specific for the test article in the mixture must be used since the vehicle matrix may interfere with the analytical response of the test article. For example, an analytical method for analyzing a test article mixed in feed will often require an extraction procedure in order to obtain a clean response on the analytical instrument. If this is the first time that the specific mixture is being used, the analytical method may have to be developed and validated, which should be taken into consideration since this will take additional time and likely add cost to the study.

Typically, whenever uniformity (homogeneity) is assessed, concentration is also assessed. Specifications for acceptable variations of concentration (e.g. ± 10% of the target concentration) and uniformity (e.g. not greater than a 5% difference between different sampling locations) should be specified in the study protocol. Homogeneity is typically assessed by collecting samples from at least three different levels of the mixture (e.g. top, middle, and bottom). When concentration and uniformity are assessed at the same time, it should be clearly indicated which sampling location is used to assess the concentration (e.g. the middle level). Samples for concentration and uniformity can be collected at any time after preparation but it may be beneficial to collect the samples shortly after dose preparation to prevent any possible degradation or external contamination of the samples.

Stability of a mixture can be determined prior to or during the study. Ideally, stability should be determined prior to the study so that you clearly know how often the mixture should be prepared and you do not run into any unexpected results that could invalidate the study (e.g. the mixture exhibits significant degradation towards the end of its use). Stability should be determined for at least the duration of use (e.g. if the mixture is prepared fresh weekly, then stability for the entire week must be shown) and under the conditions of use (e.g. if the mixture is stored at room temperature, stability at the highest expected room temperature should be tested). Mixture stability can be assessed during the study by comparing the concentration of the mixture immediately after preparation to the concentration taken at a specified time after preparation (e.g. after one week if the mixture is used for an entire week). Even if stability information is available prior to the study, it can be beneficial to conduct additional stability assessments during the study to ensure that the storage and handling procedures maintain mixture stability. As with concentration and uniformity, acceptable stability specifications should be specified in the protocol or a Standard Operating Procedure (SOP).

A final note about mixtures is that the expiration date must be listed on the container. If any mixture component has an expiration date, the earliest expiration date must be listed on the storage container if it occurs before the expiration of the mixture itself. If stability assessment of the mixture is ongoing, then the container should list that the protocol should be referenced regarding the expiration of the test article (i.e. that stability testing is being conducted concurrently with the study and any periodic analytical results should be referenced to determine the acceptability of the mixture for continued use).

TEST ARTICLE AND MIXTURE RECEIPT, STORAGE, AND TRACKING

Test articles should be sent to the laboratory using a courier that provides tracking information so that you know where the shipment is, how long it took to get to the laboratory, and who at the laboratory signed for it. If the test article is sensitive to variations in environmental conditions, you need to consider shipping the test article under controlled conditions and possibly including means to monitor the environment during the shipment (e.g. min/max thermometer). If the laboratory has the analytical capability, you can also have them analyze the test article to confirm it still meets CofA specifications after they receive it. Some of these controls may sound extreme, but it is important to ensure that you start a study with test article that is not degraded and meets the CofA specifications. The company/department shipping the test article will include various paperwork (e.g. Material Safety Data Sheet) to comply with different shipping regulations depending on the amount and type of material (e.g. hazardous, corrosive) that is being shipped. You may also want the CofA (or similar documentation) and methods of synthesis included with the shipment. Alternatively you can provide this information to the laboratory separately.

When the laboratory receives the test article, they must have a process for logging in the test article and tracking its distribution throughout the laboratory, often referred to as chain-of-custody procedures. Most laboratories will receive the test article at the shipping/receiving department. The shipping/receiving department will log in the test article according to the relevant laboratory SOP (you should review this SOP during an audit of the laboratory) and then distribute it to the formulations department for storage. The SOP should clearly outline the chain-of-custody procedures for distribution of the test article with several key items being who made the distribution, who received the material, what was distributed, and the amount distributed. Once the formulations department receives the test article, they should log in the test article and store it under the specified conditions (e.g. 4°C, room temperature). The formulations department should have separate areas for the storage of test articles, control articles, and mixtures. These areas should be secured and only accessible to authorized personnel. When the formulations department uses the test article to prepare mixtures or distributes unformulated test article, they must log at least the date of use, who removed the test article, how much they used/distributed, the purpose for use, and the amount remaining.

Chain-of-custody procedures must be used when the formulations department distributes the test article/mixtures for dosing or analytical work. The chain-of-custody procedures must document the receipt and distribution of each batch including the date and quantity of each batch distributed or returned. In addition, the distribution procedures must prevent contamination, deterioration, or damage of the test article/mixtures. When dosing or analytical work is complete, the remaining test article/mixture should be returned to the formulations department using appropriate chain-of-custody procedures. When auditing a laboratory, you should carefully review their chain-of-custody procedures to ensure that test article and mixtures are adequately tracked and accounted for.

When the study is complete, reserve samples of the test and control articles must be archived. The amount should be sufficient to conduct future analytical work to verify composition and purity, if needed. For any remaining test article, the laboratory may ask if you want it stored, destroyed, or returned. The laboratory may charge a fee for continued storage, destruction, or return shipment. If you designate destruction, they may request additional information about appropriate disposal methods or they may just dispose of it as hazardous waste. If you ask the laboratory to destroy the test article, it is best if they use a method that permanently destroys the test article and container (e.g. incineration). If the test article is disposed of in a landfill, there is a remote possibility that someone may find the test article container and raise concerns and/or disclose the information publicly.

ADDITIONAL CONSIDERATIONS

When you audit a study and ideally before the study starts, you should thoroughly understand and review how the test article will be handled, tracked, and formulated (when applicable). If special storage conditions are required, you should make this very clear to the laboratory and not assume they read all the details in the paperwork or on the container. For mixtures, the laboratory should develop a detailed formulations procedure and you should carefully review this to ensure they are formulating the test article correctly. There may be special considerations that are not adequately described in the formulations procedure. For example, you may require that the final mixture be cold filter sterilized but the formulations procedure just specifies a sterile preparation. The formulations department may assume that the test article, provided in a crimped vial, is sterile and just add sterile saline to make the final dose solution. However, the test article is not sterile and the final solution needs to be sterile filtered. These are the types of details that need to be covered in the formulations procedure.

You should also understand any sampling procedures and schedules that will be used to verify mixture concentration, homogeneity, and stability and ensure that these procedures/schedules are adequate for assessing the mixtures. You should know the accept/reject limits for concentration, homogeneity, and stability. Your specific mixture may require unique limits and you should ensure that the laboratory is aware of them. The ultimate purpose of these periodic analyses is to ensure that the animals are being exposed to protocol-specified quantities of the test article. These analyses help identify any problems that may arise during formulation (e.g. mixing equipment malfunction, out-of-calibration pipette or balance) and should be frequent enough to give you confidence that the animals are receiving the specified dosages. It is important to note that even for single dose acute studies, this information is required under GLPs. For these short-term studies, the cost of the analytical work can actually exceed that of the in-life portion of the study. Ideally, the formulation used in the acute studies can also be used in subsequent longer-term studies so that the analytical method does not have to be redeveloped and/or validated.

For mixtures, the laboratory may be formulating the test article in the specific vehicle for the first time (i.e. the vehicle was not used for any other study or this is the first study conducted with the test article). If this is the case, the laboratory will have to develop a GLP analytical method specific for the test article in the vehicle. Having an analytical method for the neat, unformulated test article is a good start and can help the analytical laboratory develop an appropriate method. Also, if you have an analytical method for the test article in the specific vehicle but the work was conducted at another laboratory, it will have to be transferred and validated at the new laboratory. An alternative approach is to have the laboratory making the formulations and conducting the dosing ship samples to the analytical laboratory that already has the method validated.

Depending on the size of the laboratory, they may or may not have analytical capabilities or may not have the specific analytical capabilities required for the formulated test article (e.g. the analytical method is a high performance liquid chromatography mass spectrometer-based method). In these cases, the formulated samples will have to be shipped to an appropriate analytical laboratory. All shipments should be made using appropriate chain-of-custody procedures. One potential issue to be aware of is that shipping conditions is another factor that may impact the test article (e.g. extreme temperatures may alter the stability). If the mixture passes specifications, then it is likely that the mixture was acceptable during dosing; however, if it fails, it is possible that the mixture was acceptable for dosing but exhibited premature degradation during shipment or other shipping-related problems. Therefore, it is best if the laboratory has analytical capabilities to avoid this potential confounder. If shipment is unavoidable, you should at least be aware of potential issues that may arise due to the additional shipment variable.

Chapter 10

Study Start Through End of In-Life

Amy Babb BS*, James Greenhaw BS, LAT*
and William F. Salminen PhD, DABT, PMP†

*National Center for Toxicological Research, FDA, Jefferson, AR, †PAREXEL International, Sarasota, FL

Key Points

- A pre-study meeting should be held to ensure that everyone understands the protocol and all of the relevant details have been covered
- The first day of dosing often entails the initial conduct of many critical study functions; therefore, it is beneficial if the Contracting Scientist is present or readily available to answer any questions
- Although the duration of the necropsy is short, it involves many critical collections; therefore, all of the details should be thoroughly worked out and for more complex necropsies, a test necropsy can be conducted to work through the logistics
- Conducting a thorough interim data audit helps identify any issues that may be occurring during the study

---

Once you have the protocol finalized you need to start preparing for the start of the study. The start of the study is not just the first day of dosing since many functions occur well before the first day of dosing and even before the acclimation period. This chapter provides an overview of the critical functions that occur throughout the study so that the Contracting Scientist can ensure that each function is conducted on time and in a quality manner. Staying on top of the study not only ensures timely conduct of the study but also helps avoid unexpected problems.

TEST ARTICLE AND FORMULATIONS

As reviewed in the previous chapter, one of the first things you need to consider is obtaining test article and any relevant documentation (e.g. Good Laboratory Practice [GLP] certificates of analyses [CofA]). You also need to determine if

Nonclinical Study Contracting and Monitoring. http://dx.doi.org/10.1016/B978-0-12-397829-5.00010-7
2013, Published by Elsevier Inc.

the test article needs to be formulated and what analytical work will have to be conducted. This process can start even before the protocol is finalized. Since the success of any study depends on administering the appropriate test article and the specified dosages, it is critical that all of the test article and formulation details be worked out before dosing starts. The one exception is stability assessment, which can occur concurrently with the study. However, it is advisable to know the stability of the test article and any formulations prior to dosing since unstable formulations could invalidate the study.

ACQUIRING ANIMALS

Before you can start a study, you need to acquire animals. Many laboratories purchase their animals from a commercial breeder. You should ensure that the breeder is known for raising quality animals specifically for research and has relevant documentation ensuring animal breeding history, genetics, health, medical treatments (e.g. vaccinations), and/or other pertinent requirements/ information. As mentioned in previous chapters, some breeders go to great lengths to ensure that their animals have calm temperaments, particularly for larger animals, and are amenable to dosing and other study procedures. This can facilitate the conduct of the study and it may be worth purchasing animals from these types of breeders even if the animals cost more. The laboratory may have its own breeding colony that will supply the animals for the study. If this is the case, you need to ensure that they have the same breeding controls, procedures, and documentation as any other commercial breeder. If you have to source animals from a breeder that does not specialize in research animals (e.g. you require a specific strain or breed of animal) you need to conduct a more in-depth audit of the breeder's procedures and documentation to ensure that you will obtain animals that meet the study requirements and comply with applicable animal welfare regulations.

If the animals are being sourced from an off-site breeder, they will need to be shipped to the laboratory. The breeder must follow appropriate regulations for shipping the animals and should ensure that the animals are exposed to minimal variations in environment (e.g. animals should be shipped in a temperature controlled truck) to minimize shipping stress. The animals should be provided with appropriate feed and water for the duration of the shipment. Once the animals arrive at the laboratory, they should be placed in an animal room for quarantine until their health status can be assessed by a staff veterinarian. It is best if the animals arrive at the laboratory a couple of weeks before the study starts to ensure that they are healthy and fully acclimated to the study conditions before dosing starts. However, since maintaining animals at a laboratory ties up animal space, it is not uncommon for laboratories to keep the duration between animal receipt and the start of dosing as short as possible. It is important to determine what period is needed to ensure adequate acclimation to the laboratory environment. It may be necessary to negotiate or pay extra for a prolonged holding period (e.g. you receive very young animals that are more likely to be subject to shipping stress and require a longer acclimation period).

Laboratories may have different procedures for separating or defining quarantine, holding, acclimation, and/or baseline periods. The important point is that when the animals are received, they should be held in an isolated room until their health status can be assessed. This can be the same animal room they are housed in for the study if no other studies are being conducted in the same room. Once their health status is verified, they can be released from quarantine and used in the study. The acclimation period allows the animals to adapt to the laboratory environment and may overlap with the quarantine period. Some laboratories may also incorporate a baseline period that occurs after the acclimation period. Regardless of whether or not the various periods overlap and what terminology is used, you need to ensure that the animals are held for a sufficient length of time to allow them to adapt to the laboratory conditions before dosing starts. In addition, if baseline measurements are taken during the acclimation or baseline periods, you want to ensure that the baseline measurements are taken after the animals are fully acclimated to the laboratory environment. This may entail using a prolonged acclimation period with baseline measurements taken towards the end of the period after the animals have fully acclimated.

As described above, the health of the animals must be assessed by a veterinarian before they are released from quarantine. It is also best if the animals are checked again prior to enrollment in the study. This may be a simple health status check or a complete physical examination. It is best if it is conducted shortly before the animals are assigned to groups (e.g. towards the end of the acclimation period) so that you are assured that only animals meeting the study requirements are enrolled in the study. Once animals are assigned to groups, you need to decide what to do with any extra animals you may have ordered. It is often best to hold these animals in the same room for a period in case they are needed for the study (e.g. a dosing error occurs on the first day of dosing resulting in an animal's death and one of the extra animals can be enrolled in place of the dead animal).

PRE-STUDY MEETING

Unless the study is very routine and simple, most laboratories will hold a pre-study meeting to discuss the study details. Personnel representing all of the relevant laboratory functions (e.g. formulations, in-life, clinical pathology, pathology) will come to the meeting prepared with any questions they have about the protocol/study. The Study Director should have worked out many of the kinks with the draft protocol and solicited feedback from the different groups prior to the meeting; however, the individual groups may not have had an opportunity to read the final protocol until shortly before the meeting. They may find different logistical issues with the protocol or may need to work out details of how different functions will be conducted or handed off between groups.

This is a great time to allow everyone to raise questions and work out the study details since the Study Director may have solicited individual feedback but it isn't until everyone is in the same room and talking about the details that potential issues become apparent. An easy way to go through the protocol is for the Study Director to run through each section of the protocol, highlight any problems they anticipate, and solicit input from each function.

It is often beneficial if the Contracting Scientist participates in the pre-study meeting. This can be in person or via teleconference. This gives the laboratory a chance to ask any relevant questions of the Contracting Scientist that the Study Director may not be able definitively to answer. Also, if the protocol needs to be modified, the Contracting Scientist can provide their input and give their approval or request a different type of change that better meets the needs of the study. This open discussion can greatly reduce the time required to finalize the protocol since issues can be discussed and hopefully resolved at the pre-study meeting instead of the Study Director having to relay the information after the meeting. Attending the pre-study meeting also sends a clear message to the laboratory that the Contracting Scientist will be actively involved with monitoring the study and expects a well-conducted, high quality study.

The protocol may or may not be finalized before the pre-study meeting. If the protocol is still in draft format, it should at least be in a final draft format that is as close to the final protocol as possible. This allows each laboratory function to assess accurately what is needed for the study and how the study will be coordinated and conducted. If changes are required to a draft protocol, the changes can be made without an amendment. Once the changes are made, the protocol needs to be reviewed and approved according to the GLPs and laboratory Standard Operating Procedures (SOPs). If the protocol has been finalized, then a protocol amendment will need to be made covering the change(s). The amendment needs to be reviewed and approved using the same process as for the full protocol. Ideally, the Study Director has worked out all of the key protocol details prior to the pre-study meeting and only minor details need to be discussed that do not impact the wording of the protocol and do not require a protocol amendment.

FIRST DAY OF DOSING

Many different study functions can occur prior to the first day of dosing, especially if baseline measurements (e.g. clinical pathology samples, detailed physical examinations) are collected. However, the first day of dosing is often a critical event since that is the first day that animals are exposed to the test article. The first day of dosing often has other study functions beginning and can be quite hectic for complex studies. Therefore, it is often beneficial for the Contracting Scientist to be physically present at the laboratory for the first day of dosing. At the very least, the Contracting Scientist should be readily available via phone to answer any questions or address any problems that may arise. Thorough planning for the study and working through the study details helps avoid the majority of problems; however, issues can still arise (e.g. difficulty dosing a viscous oral solution, timing deviations due to tightly spaced pharmacokinetic sampling intervals interfering with the conduct of other study functions). Being present during the first day of dosing when many of these functions are starting allows the Contracting Scientist to work with the Study Director and other laboratory personnel to address the issues, plot a course of action, and make any necessary changes to the protocol and study procedures.

It is often beneficial to arrive at the laboratory one or two days prior to the first day of dosing. This allows you to review the animal procurement paperwork, health checks, formulation information/results, and any data generated to date. You can also make sure the Study Director and everyone else participating in the study is prepared for the first day of dosing and the timeline of events flows well. For example, when will the formulations be prepared, when will they be transferred to the in-life group, what time will dosing start, when will clinical observations be conducted, when will any blood draws be taken and what is the procedure (e.g. peripheral vein stick for each draw). If there are unique functions that occur on the first day of dosing, it is best to go through these with the Study Director and relevant personnel to ensure they understand what needs to be done and how it will be coordinated with the other study functions.

You should ask the Study Director when you should arrive at the laboratory on the first day of dosing. Many laboratories perform dosing early in the morning and you should arrive in plenty of time prior to the dose since you will need to sign in, don protective gear, and be escorted to the animal room. If mixtures are prepared that day, you may even want to watch the formulation preparation, labeling, and transfer, which would occur well before dosing starts. You should bring a notebook and a copy of the study protocol with you so that you can reference it when making observations. Even though you should be well versed in the protocol details, it is always handy to have a copy in case you need to confirm specific details.

When you are observing dosing, it is important to note if the right formulations are being administered to the right animals. You should verify this by noting the formulation information on each container and recording which animals it was administered to. If there are critical time windows for dose administration, you should also note the times to ensure that the timeframes were met. When other study functions are conducted, you should make a note of when they were conducted (e.g. weighed-in feed weights were measured between 8:00 and 8:30 a.m. for all animals, clinical observations were conducted on all animals between 10:00 and 10:30 a.m.) and if they complied with the protocol requirements. For critical timed functions (e.g. closely spaced pharmacokinetic blood draws), you will want to pay very close attention to the time windows. Throughout the day, you should ensure that all protocol-required study functions are conducted. In addition, it is important to make sure that extra, non-protocol study functions are *not* conducted. You should note how data are being recorded and ensure that it is accurate (e.g. does the actual dose administration time match the computer or paper records, are all clinical observations accurately recorded).

At the end of the day or the following day, it is useful to review the data collected on the first day of dosing. If data are collected via computer systems, you will need the Study Director to make relevant printouts and/or show you how to access the data electronically. For data collected on paper, you will need to obtain the data collection forms from the first day. You should review the data to make sure all protocol-specified functions were conducted and accurately recorded and data are being collected in compliance with GLPs. You should make sure that all relevant time windows were met. If you find inconsistencies or protocol deviations, you must notify the Study Director. The Study Director can then have the inconsistencies corrected or clarified and the deviations written as formal protocol deviations. The deviation should clearly state the issue and what will be done to prevent further deviations. A protocol amendment may be needed to adjust a procedure so that further deviations do not occur.

PROTOCOL DEVIATIONS AND AMENDMENTS

As the study progresses, particularly for longer studies with many different study functions, there are likely to be instances where various functions were not conducted according to the protocol- or SOP-driven specifications. For example, the protocol specifies that blood draws were to be taken 15 ± 5 minutes after the dose and some were collected at 21 minutes after the dose. A more serious deviation is if the protocol and/or SOP specifies that health, water, and feed checks are supposed to be conducted twice a day but a technician forgot to check the animals during the weekend due to a scheduling error. These are protocol deviations since they were not planned. Any deviation needs to be adequately documented, a corrective course of action determined to prevent similar deviations, and all relevant personnel made aware of the issue and corrective action. It is important for the Contracting Scientist to review and understand the laboratory's procedure for recording, addressing, and communicating deviations and the corrective actions. For example, who records the deviation, who identifies the corrective action(s), how soon is the Study Director notified of the deviation and how is this documented, and how is the deviation/corrective action circulated/communicated to laboratory personnel. Since the Study Director is the single point of control for the study, it is critical that the Study Director be immediately notified of all deviations, no matter how minor, so that they can assess the impact on the study and determine corrective actions. A deviation may seem minor to a person conducting a certain function but they may not have a complete understanding of how it might impact the overall study and it is up to the Study Director to determine the overall impact since they are the single point of control for the study. The Study Director should in turn notify the Contracting Scientist of deviations in a timely manner, especially ones that have a potential impact on study integrity. Before a study starts, you should clarify with the Study Director the timeline for reporting deviations and if different types of deviations (e.g. minor versus serious) will have different reporting requirements. It is not unreasonable to request that as soon as the Study Director receives a deviation, they send a copy of the deviation to the Contracting Scientist for review, especially if you are working with a laboratory for the first time and you want to keep a close eye on the progress of the study. The Study Director must play a strong role in ensuring that deviations are kept to a minimum and that study personnel are fully aware of any issues as the study progresses so that future deviations are avoided. An excessive amount of deviations not only makes the study look like it was poorly conducted, but if the deviations are serious, they could invalidate the study. Thorough planning of the study is the best way to avoid deviations; however, they can still occur due to unforeseen circumstances. The key is to ensure that the deviations are identified as soon as possible, the Study Director is immediately notified, and procedures are implemented to prevent further similar deviations.

After the protocol is finalized, it may need to be modified. This could be simply to add a study start date or to add a completely new procedure. If there are many protocol deviations due to a given procedure, the procedure may be altered via a protocol amendment to prevent further deviations (e.g. the time window for completing detailed behavior assessments are too tight). In the case of trying to avoid future deviations, it is important to determine if the change can be made without altering the scientific integrity of the study or if other approaches (e.g. adding manpower), besides modifying the protocol, are needed to avoid the deviations. The protocol is modified by making a protocol amendment. The protocol amendment should clearly state what is being changed or added and the reason for the change/addition. The protocol amendment must be reviewed and approved in the same manner as the protocol (e.g. if Test Facility Management and the Sponsor signed the protocol as an acknowledgement that they reviewed and approved the protocol, then they should sign each amendment). Under GLPs, the Study Director has to review, approve, and sign/date all protocol amendments. Copies of the amendment must be placed with each copy of the protocol for easy access. If the amendment is not placed with each copy of the protocol, then the personnel reading the protocol have no way of knowing that an amendment was issued for that protocol and what the change entails. The laboratory should send you a copy of each amendment so that you can keep it with your copy of the protocol. The laboratory must have a defined process for ensuring that all study personnel are aware of and understand each protocol amendment that applies to studies they are involved in. The Study Director must take a lead role in this process since they are the single point of control for the study. You should review and clearly understand the laboratory's protocol amendment process to ensure that it meets your requirements and results in thorough notification of all relevant study personnel. For laboratories that have poorly defined notification procedures, it is not uncommon for an amendment to be issued during the week, a weekend technician is not notified of the amendment (e.g. their supervisor failed to tell them about the amendment and/or a copy of the amendment is not placed with the protocol), and the technician fails to conduct a procedure according to the amendment resulting in a protocol deviation.

STUDY UPDATES, DATA AUDITING, AND GLP COMPLIANCE

Before the study starts and possibly even within the contract, you should determine how often the Study Director will provide you with study updates and how extensive they will be. The frequency may depend on the length of the study or the timing of critical study functions but should be frequent enough to give you confidence that the study is progressing as planned. It is useful if the Study Director can send you summaries of the data collected to date. For example, they can send you feed consumption and body weight data, clinical pathology results, reproduction performance results, to name a few. You should clearly specify up front the detail you require. For example, is a simple "there are no test article-related changes to date" acceptable, do you require group means and standard deviations, or do you require clinical pathology values for each individual animal. It is important to hold the Study Director accountable for providing these updates on time since they are an important part of ensuring the study is running well. You will also want the Study Director to notify you immediately of any deaths, moribund animals, or other serious unexpected events. Although this seems like common sense, it is important to set this expectation before the study starts so that there are no misunderstandings.

Although the Study Director and Quality Assurance Unit (QAU) should thoroughly monitor the study to ensure that data are being recorded and the study is being conducted in compliance with GLPs, it is important for the Contracting Scientist to conduct an independent review of the data. Due to high workloads and the sheer volume of data in some studies, the Study Director and QAU may not catch every inconsistency or error. Also, some laboratories may simply not have a strong focus on data review. As mentioned previously, data auditing should start at the beginning of the study, ideally prior to, during, and/or shortly after the first dose. In addition, especially for longer-term studies, you may want to conduct interim data audits. Often, the easiest way to conduct these is by visiting the laboratory and reviewing the paperwork and data collected to date. Some laboratories can also provide electronic access to study data via secure portals so you can review the electronic data from your office.

You should review the paperwork and data as you would review the data in a draft report. A systematic process for reviewing study data accompanying a draft report is presented in Chapter 13. A similar process can be used to ensure that the protocol and SOPs were followed and the data were recorded in compliance with GLPs. The major difference with an interim audit is that you should also review any paperwork or data that may not be incorporated into the draft report (e.g. animal receipt records, animal release records, animal room environment records, water analysis results, timing of events such as dosing and clinical observations, QAU findings [the laboratory should let you see these even though they do not have to let FDA inspectors see them]). Conducting a thorough audit and in-depth tracing of several animals from receipt until the current study date is a great way to identify any systematic issues and have them resolved.

It has been touched upon in previous chapters that electronic data capture systems can help ensure that the right study functions are conducted, the correct animals are selected, and data are accurately recorded. However, electronic data capture systems cannot prevent all mistakes and especially cannot circumvent errors due to users disregarding warnings or not following procedures. You should have the Study Director print off data that allow you to assess technician compliance to protocol-specified procedures and identify if a technician systematically entered data that could not logistically match what was actually done. For example, you may be present for the first day of dosing and observe that the time between dosing each animal takes approximately five minutes. This duration is due to having to remove each animal from the cage, entering the animal number in the data capture system, conducting pre-dose clinical observations according to the protocol, entering the dose mixture number, administering the dose, and acknowledging in the system that the dose was successfully administered. This sequence is repeated for each animal. During an interim audit, you ask for dose timing records. You notice that on certain days and for a certain technician the duration between doses is 30 seconds. In addition, you notice that the clinical observations on these days match exactly those from the previous day. There is no way the doses could have been administered this quickly and it is very odd for all of the clinical observations to match from day to day. It is likely that the technician systematically entered data in the system for every animal either before or after they conducted the actual dosing (i.e. they went through all the system prompts repeatedly for every animal and once this was completed, they proceeded to dose the animals). They also most likely did not conduct the specified clinical observations. These are major GLP deviations and it is even possible that the technician may never have dosed the animals. By looking at only summary data and not reviewing the study data in detail, you would not be able to identify these types of issues. That is why it is critical that you delve into the data and conduct a thorough audit of all of the data. The Study Director may balk at having to print out these detailed reports; however, you need be assured that your study is being well conducted and the only way to confirm this is by thoroughly reviewing the data. These types of audits are not restricted to electronically captured data and also apply to data recorded on paper. If paper is used, it is actually easier for the Study Director since they can just bring you the relevant binders instead of having to print out detailed reports.

If you do not find any major issues, you will be reassured that the study is progressing as planned and the data accurately represent what has been done to date. This lays a solid foundation for completing the study and obtaining a draft report that reflects the actual conduct of the study and contains accurate data. Once the study is completed, you may want to conduct a final close-out audit. This final audit follows the same process as an interim audit and ensures that all the paperwork and data are in order and will pass any future regulatory audits.

If you visit the laboratory to conduct an interim data audit, you should also audit the study procedures being conducted (e.g. dosing, clinical observations, blood collections) to ensure that the laboratory is still remaining compliant with the protocol and the study is progressing as planned. If you have unique study functions being conducted (e.g. semen collection and analysis, eye injections and examinations by a veterinary ophthalmologist, echocardiograms), you may want to time your visit to coincide with those functions so that you can make sure they are performed according to the protocol and answer any last minute questions that might arise.

NECROPSY

If your study includes a necropsy, you need to make sure everyone is prepared for this critical event. A necropsy is a small part of the study in terms of the absolute amount of time involved (e.g. one or two days); however, there are many critical functions and collections that need to be made during this short time period. Also, the hand-off between the in-life group and the necropsy group needs to be clearly outlined and coordinated. For example, do animals need to be fasted prior to necropsy, when will animals be delivered to the necropsy laboratory, who will conduct the final body weights, and who will conduct terminal blood draws, just to name a few. A necropsy can range from a simple gross necropsy with observation and collection of a single organ to a very detailed necropsy with collection of every major organ and tissue, special preservation techniques (e.g. flash frozen tissue, RNA-specific preservatives, special blood collection tubes and processing procedures), and many organ weights. Depending on the complexity of the necropsy, it can be beneficial to hold a pre-necropsy meeting with relevant laboratory functions (e.g. in-life, clinical pathology, necropsy). As with a pre-study meeting, the Contracting Scientist may want to participate in the meeting so that they make sure the necropsy will go as planned and can answer any questions. The necropsy group may develop a necropsy procedure memo that clearly outlines the observation and collection procedures. This procedure memo can be included as an appendix to the study protocol or it may just be included in the raw data. The procedure memo is especially beneficial for necropsies that involve unique procedures (e.g. tissue collection for mRNA expression that requires special decontamination, handling, and preservation procedures to avoid RNAase contamination) since it provides a detailed sequence of events for each technician to follow. This procedure memo should be reviewed by the Contracting Scientist to ensure that it meets the scientific needs of the study. If the necropsy involves a lot of procedures that may lead to logistical issues (e.g. blood and tissues need to be collected and preserved within a certain timeframe after euthanasia) or the necropsy involves procedures/collections that the laboratory does not routinely conduct, it can be helpful to run a test necropsy on several non-study animals (e.g. extra animals that you ordered but did not enroll in the study) using the exact same number of personnel per animal and following the same procedures that you will use for the final necropsy. This test necropsy can make sure that the procedures run smoothly and if any issues arise, you can resolve them prior to the final necropsy. If you will be conducting any procedures on the test animals prior to euthanasia (e.g. blood collection), you will need to obtain specific IACUC approval if these animals were not included in the original protocol.

Study personnel are critical to the success of any necropsy. You should ensure that all of the technicians involved with the necropsy are not only thoroughly trained, but they have sufficient experience with the given species and procedures, whenever possible. You should avoid having a necropsy technician assigned to your study that just completed their training since they are likely to be much slower than other experienced technicians, may have to ask a lot of clarifying questions, and may make more mistakes. Also, they may have difficulty conducting some procedures (e.g. conducting terminal blood draws, dissecting an organ in a timely manner) resulting in inadequate sample for analysis or missing critical time windows.

The protocol should specify who will supervise the necropsy. It is often beneficial if a board certified veterinary pathologist not only supervises the necropsy but is physically present in the laboratory during the necropsy. This allows them to observe readily any gross lesions and answer any questions in a timely manner. A board certified veterinary pathologist has extensive training and experience in veterinary pathology and is the most qualified person for supervising a necropsy. The only downside to using a veterinary pathologist is that the laboratory may charge a premium, especially if they have to be present in the laboratory during the entire necropsy, for their supervision of the necropsy. For some specialized studies (e.g. reproductive or developmental studies), it may be beneficial to have a highly experienced technician or scientist co-supervise the necropsy since they may be more familiar with the specific procedures and possible findings.

Chapter 11

Study Communication and Data Management

William F. Salminen PhD, DABT, PMP* and Xi Yang PhD†

*PAREXEL International, Sarasota, FL, †National Center for Toxicological Research, FDA, Jefferson, AR

Key Points

- Clear lines of study communication are critical to the success of any study and these expectations should be set before the study starts
- Issues often arise during a study and you should keep a calm head, thoroughly understand the problem, and then chart a course of action
- It is important to have a logical system for filing and archiving studies at the Sponsor's company

Study communication and data management are critical to the success of any study and ensuring that you can easily keep track of the study as it progresses. The previous chapter already covered the importance of setting clear expectations with the Study Director for the communication of regular study updates, deviations, unexpected adverse events, and deaths/moribund animals. Without timely notification, you will not be able to keep abreast of the study and make any necessary changes to maintain the integrity of the study and meet your scientific requirements. Therefore, you should not only set clear expectations, but also hold the Study Director and laboratory accountable for meeting the pre-specified timelines. For example, if the Study Director is to notify you immediately of dead animals, as soon as they are notified by the technicians or animal care staff of the animal death, they must make reasonable efforts to contact you. If they cannot reach you by one method (e.g. cell phone), they should try others (e.g. work phone, e-mail) or contact a back-up person, if such a person is provided in the protocol. At the very least, they need to leave a message (e.g. voice mail) about the situation so that you can contact them as soon as you receive it. If you will not be available by the means listed in the protocol (e.g. you will be traveling in a remote location without cell phone service), you should notify the Study Director and provide a back-up contact who can make decisions about the study in your absence.

Nonclinical Study Contracting and Monitoring. http://dx.doi.org/10.1016/B978-0-12-397829-5.00011-9
2013, Published by Elsevier Inc.

The Study Director may have to take an absence from the study, such as for extended travel or vacation. In these cases, the laboratory should assign another Study Director to the study so that they can keep adequate oversight of the study. Depending on the laboratory's Standard Operating Procedures (SOPs), a protocol amendment may be issued to change the Study Director for the given time period. When the laboratory changes the Study Director, you should ensure that not only do they fully understand the protocol and any unique functions but that they continue to meet the communication expectations that you set with the original Study Director. Since the new Study Director will have an extra workload placed on them, it is particularly important that you emphasize that they must still continue to provide thorough oversight of the study and meet your communication requirements.

HANDLING ISSUES THAT MAY ARISE

There is rarely a study that is completed without some unexpected event. Many of these are unforeseen and are not the particular fault of any one party. Sometimes the wording in the protocol can trip up a study resulting in deviations, key study personnel may not be available as scheduled (e.g. a consulting veterinary ophthalmologist is sick and cannot make a scheduled examination), ongoing test article mixture analysis may show poor homogeneity despite previous formulations being stable and homogeneous, among many others. In any situation that does not go as planned, it is important that clear lines of communication are available not only to convey the issue in a timely manner but also determine a corrective course of action. As mentioned above, you should establish the communication expectations before the study starts and make sure everyone has each other's emergency contact information (e.g. cell phone number). When an issue arises, you should take time clearly to understand and assess the situation. You want to act quickly to avoid perpetuating the situation but you also do not want to be hasty in your decision until you have all the facts surrounding the issue. It is important to make sure that the Study Director conveys all of the necessary details including what is the issue, why did it happen, how does it affect the study integrity (this may be a question that only the Contracting Scientist can determine), and what corrective course of action do they recommend. The immediate concern lies in correcting the problem so that it does not occur again. Once you have this information, take time to thoroughly review and understand the information and ask any clarifying questions. You may even want to consult with other scientists or relevant experts (e.g. company Quality Assurance Unit [QAU]) about the impact on the study and potential corrective courses of action. After you make a final decision, make sure that the Study Director implements any necessary changes as quickly as possible so that any similar problems do not recur. Some issues may not require a corrective course of action, but you should still make an assessment about how the issue may affect the overall integrity of the study. For example, if the sick ophthalmologist had to delay the eye examinations by several days, does this invalidate the study?

After you have assessed the situation and determined a corrective course of action, you can then determine if a certain party is at fault. The vast majority of issues are not due to negligence or incompetence and cannot be blamed on a single party. However, situations do arise where the laboratory is clearly at fault. For example, an inexperienced formulations technician prepared test article mixtures using 400 mg of test article instead of the required 4000 mg since they did not know how to use and read a particular balance, an in-life technician administered 5 ml/kg of a test article formulation instead of the required 10 ml/kg, or a necropsy technician failed to open the gastrointestinal tract to observe ulcerations despite this being clearly listed in the protocol. When these types of situations arise, you need to determine how serious the issue is and if the integrity of the study is so compromised that it will have to be reconducted. When the situation is clearly the fault of the laboratory and there is no question that the study will have to be reconducted, most laboratories will not hesitate to offer to reconduct the study as soon as possible. Since your project timelines may hinge on the given study, you should not hesitate to ask the laboratory to provide you with a discount on the original study cost to offset any additional costs your company may incur due to the project delay (e.g. a delay in the time to market can result in a significant decrease in earnings for a number of different reasons – you can ask your company marketing experts for help in determining the cost of the delay). Do not expect to recoup the full cost of the delay since it is likely to be much greater than the cost of the entire study; however, you should at least receive some reduction in the study cost as a good will gesture. The most difficult situations arise when there is a disagreement on who is at fault or if the issue truly has an impact on the integrity of the study. The laboratory may agree that they are at fault but insist that the problem does not impact the integrity of the study. For these situations, it is best not only to discuss the issue with the Study Director but also Test Facility Management since they have the power to make decisions about reconducting the study. Depending on how initial talks go, you may need to involve your legal council. You may even want to do this during your first discussions with Test Facility Management. It is best if the issue can be resolved to the satisfaction of all parties without having to resort to legal means; however, some situations may require a legal course of action and that is where having a solid contract can help resolve any disputes. Fortunately, most laboratories realize that a large portion of their business is due to repeat business and peer-to-peer recommendations; therefore, they will often seek a compromise that appeases the client (i.e. Contracting Scientist).

DATA MANAGEMENT

Data management is important not only at the laboratory but also at the Contracting Scientist's company. The aspects of data management for the laboratory have already been covered in previous chapters. This section will cover data management at the Contracting Scientist's company.

Although the Good Laboratory Practice (GLP)-compliant data collected during the study will be at the laboratory when the study is underway, you will be receiving copies of various types of data so that you can keep track of the study (e.g. interim updates from the Study Director). You may also have copies of other information related to the study such as test article sourcing and shipment documents. It is ideal if you can establish a system for filing copies of data for each study so that it can be easily retrieved and reviewed by authorized personnel. For example, for paper copies, you could establish study-specific binders in an archive room. For the electronic data, you can establish study-specific electronic folders or databases where the electronic data are filed. Access to the data should be restricted to ensure that only necessary personnel have access to the files. This is to prevent unauthorized access and keeping people from removing or deleting data without appropriate controls. Many projects have confidential information and the fewer people that have access to the information, the less likely it is to leak from the company. For paper data, having a sign-in/sign-out system is best so that data can be accounted for. The filing system used for interim data can also be used to file the eventual final study report, which will be covered in a later chapter. For electronic data, restricted authorized user access controls should be used and the data should be stored in a format where only restricted individuals can delete data (e.g. the system administrator).

As you receive copies of data, you should file them as soon as feasible. If you need access to the data and they are in paper format, you may want to file them in the study binder and keep paper copies for your reference or you may want to sign out the binder. For electronic data, you can upload the files into the relevant system and access them from there. You could also keep separate personal copies of the electronic data files. If you take this approach, you should establish a filing system that allows you easily to keep track and access the data and the files should only be accessible to you (e.g. on your personal computer or a personal network drive).

In addition to copies of data, you should keep copies of all study-related communications. These can be communications with the laboratory (e.g. Study Director), within your company, or with other parties (e.g. consultants). You can keep these electronically (e.g. e-mails) and/or in paper format but they should be filed logically (e.g. chronologically) and easily searched. For phone calls discussing aspects of the study, it is best to write and file a phone record of the conversation. When you audit a study (e.g. visit for first day of dosing, interim or close-out data audits), you should write up and file a summary of your visit and findings. An example study audit template is presented at the end of the chapter. This template can be easily modified for use as a phone call log since it provides all of the essential study-specific information.

STUDY REPORTS AND ARCHIVING

The following chapters will cover the review of study reports and communication of changes to the laboratory. This section covers the receipt and filing of the reports. After the study is completed, the Study Director, with the help of other groups at the laboratory, such as a report writing group, will compile all of the data and any contributing scientist reports (i.e. sub-reports prepared by other scientists involved with the study but working at other locations such as pathology work contracted to another laboratory) into a draft report. Depending on the study contract and your preferences, the laboratory may send you paper and/or electronic copies of the draft report. Regardless of the format, the laboratory should send the reports in a secure manner that can be tracked (e.g. courier with tracking abilities for paper reports or electronic reports on media [e.g. CD-ROM] and secure e-mails or portals for web-based transfer of electronic reports). Once you receive the draft report, you should make sure it contains all of the necessary information and parts. For example, if the report requires the submission of copies of raw data, the draft report should contain the raw data copies so that they can be reviewed. Once you are reassured that the report is complete, you should log in and file the report in your document tracking system, whether it is paper and/or electronic. If the laboratory provided you with multiple paper copies of the report, you can log one in and use the other for review and mark-up. However, as will be discussed in the chapter on reviewing the draft report, it is often easiest to mark-up an electronic version of the report since this facilitates the communication of any changes with the laboratory. If you work with an electronic copy, you should not make comments in the original draft report files and instead use a working copy that you will eventually transmit to the laboratory. You should modify the filename accordingly to provide an adequate description so that you do not get the modified file confused with the original version or other versions that may be present. Since a lot of effort goes into commenting on a draft report, you may want to keep multiple versions of the file as you progress with your review. For example, you could save a new version at specified intervals. For example, draft study report "Toxicology Study #ABC-123- Draft Report – Working Copy v1.doc" would go to "…v2.doc" after you completed the review of the body of the report and then to "…v3.doc" after you completed review of the appendices. If your working file gets corrupted, you will at least be able to go back to the last version. You can also have some programs such as Microsoft Word automatically save back-up versions of the file for you. If you decide to mark-up a paper version of the report, you will need to communicate your changes to the laboratory. You can do this by subsequently marking up an electronic file, sending your handwritten mark-ups to the laboratory, or communicating your changes via phone.

Depending on the quality of the draft report and amount of changes you require, you may receive multiple versions of the draft report. You should file each version and ensure that you can clearly distinguish the different versions. Once you are satisfied that the draft report meets your expectations, you will tell the laboratory to finalize the report. After it is finalized, they will send you paper and/or electronic copies. Once you receive the report, you should check it thoroughly to make sure the laboratory made all of the agreed-upon changes and it is complete since it is this copy that will be used for regulatory submissions. You should not assume that the final report is ready-to-go for regulatory submissions. It is rare, but "final reports" can have missing sections, sub-reports, or data. You need to ensure that the final report you file in your archives is complete and ready for any submissions. Once you are assured that the final report can be filed, it is often best to discard any draft reports to avoid selecting an incorrect version for future uses such as writing toxicology summaries or regulatory submissions. Sometimes the final report has to be amended to correct errors or add information. The amendment needs to state clearly what information in the final report is being revised and what the new information is. The amendment needs to be reviewed and approved in the same manner as the final report. Once you receive the amendment, you need to make sure it is placed with each copy of the final report and it is very clear to anyone that reads the final report that it has been amended. For example, you could place the amendment at the beginning of the final report so that the reader knows an amendment has been made.

Once the study is completed and the final report has been issued, the laboratory needs to archive samples of the test and control articles, protocol, study-related information such as data and communications, the report, and any specimens. The specimens need to be retained for as long as they are still useful (e.g. not degraded). Examples include formalin-fixed tissues, hematoxylin/eosin-stained slides for pathology assessment, and pharmacokinetic samples. The study contract will often include a defined period of archiving at the laboratory (e.g. 1 year from issuance of the final report). After the initial contract period, you will need to pay the laboratory for continued archiving, have the material transferred to another commercial archive, or archive the material at your own company. It is often easiest if you pay for continued storage at the laboratory; however, some laboratories may have limited space and may charge higher than normal rates for continued storage. This is where transfer to another commercial archive can be useful. You can also archive the material at your own company; however, you should have an established GLP-compliant archive. If you decide to transfer the material to another archive, it is critical that the material is transferred by a secure, reliable, and trackable courier. The last thing you want is to have your study-related material lost in transit. If you have perishable specimens (e.g. frozen samples for pharmacokinetic analysis), it is important that you make appropriate accommodations to maintain sample integrity during shipment and after receipt.

COMMUNICATING STUDY RESULTS

Before the study starts, during the study, and after it is completed, you may be asked to give study overviews and updates. These can entail reviews of the study design, progress reports, adherence to the project timelines, and results. You may be presenting this information to internal (e.g. other scientists, project managers, marketing department) and external (e.g. regulatory authorities) stakeholders. You may discuss a single study or may give an overview of a battery of studies. These overviews and updates may occur via face-to-face meetings, teleconferences, videoconferences, and/or web-based meetings. You may also be providing updates via e-mail. Regardless of the medium, it is useful to provide some basic information in a logical manner so that the audience can clearly understand the study design, the results, the implications of the findings, and any issues that may need to be resolved. Figures 11.1 to 11.10 are presentation templates that provide this basic information. The templates are filled in with a fictional dog study to help illustrate the different types of information that is helpful to present.

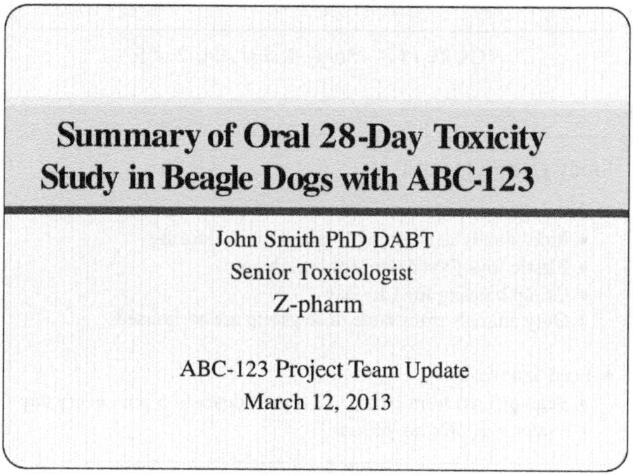

FIGURE 11.1 Title slide

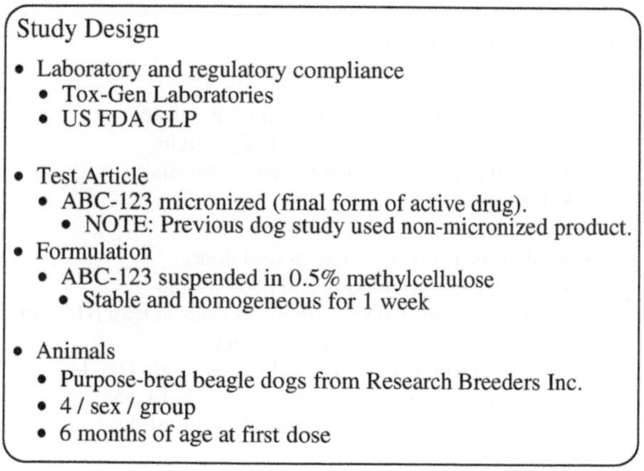

FIGURE 11.2 Study design slide 1 of 5

Study Design (cont'd)

- Route and duration of exposure
  - Oral gavage
  - Once daily for 28 consecutive days
    - All doses administered to fasted animals in the a.m.
    - Feed offered back 2 h post-dosing

- Dose groups
  - Control (0.5% methylcellulose)
  - 10 mg/kg
  - 30 mg/kg
  - 100 mg/kg

FIGURE 11.3   Study design slide 2 of 5

Study Design (cont'd)

- Housing
  - Individually in cages with raised mesh flooring
  - Plastic toys (Nylabone) for enrichment
  - Group housing for 1 h / day
  - Only animals from same dose group are co-housed

- Feed and water
  - Feed offered from 10 a.m. (2 h post-dose) to 5 p.m. every day
  - Water available *ad libitum*

FIGURE 11.4   Study design slide 3 of 5

Study Design (cont'd)

- Study Assessments
  - Acclimation period-1 week prior to first dose
    - Daily feed consumption and body weight
    - Baseline clinical pathology and urinalysis
    - Physical examination

  - Study Day 1 (first dose) to 28 (last dose)
    - Daily feed consumption and body weight
    - Detailed clinical observations 2X / day at least 6 h apart
      - First observation 1 h post-dosing
    - Clinical pathology and urinalysis on Study Day 14
    - Exposure assessment: Study Days 1, 7, 14, and 28
      - Blood collected 1, 2, 4, 8, and 24 h post-dose

FIGURE 11.5   Study design slide 4 of 5

Study Design (cont'd)

- Study Assessments (cont'd)
  - Day of necropsy (Study Day 29)
    - Animals fasted as with previous days
    - Final detailed clinical observation in the a.m.
    - Physical examination in the a.m.
    - Clinical pathology and urinalysis
    - Gross necropsy
    - Organ weights (all typical organs)
    - Histopathology (all typical organs)
      - NOTE: for liver, sections from median lobe preserved
    - Special collections
      - Sections of median lobe of liver flash frozen in liquid nitrogen for subsequent Oil-Red-O staining.

FIGURE 11.6   Study design slide 5 of 5

Results to Date
- In-life phase has been completed
- Histopathology is in progress
- Exposure assessment samples are being analyzed

- No adverse effects on:
  - Body weight
  - Feed consumption
  - Clinical observations
  - Physical examinations
  - Hematology and urinalysis parameters
  - Gross necropsy

- Treatment-related effects on:
  - Clinical chemistry- serum liver enzymes
  - Liver weight

FIGURE 11.7   Results slide 1 of 2

Results to Date (cont'd)
- Serum liver enzyme and liver weight changes on SD 29

Dose Group	Alanine Aminotransferase (IU/L)	Aspartate Aminotransferase (IU/L)	Liver Weight (g)
Control	31 ± 8	50 ± 9	480 ± 91
10 mg/kg	39 ± 11	45 ± 12	461 ± 87
30 mg/kg	73 ± 40	130 ± 25	520 ± 121
100 mg/kg	1030 ± 329	611 ± 119	680 ± 191

- All other clinical chemistry parameters were normal
- Slight trend towards increased serum liver enzymes on SD 14

FIGURE 11.8   Results slide 2 of 2

Conclusions
- ABC-123 induced dose-dependent liver toxicity based on serum liver enzyme levels and increased liver weight.

- Histopathology of liver will be assessed for all dose groups to confirm findings.

- ABC-123 in previous 7-day dog study induced signs of fatty liver but no other signs of liver toxicity.
  - Oil-Red-O staining for lipids is pending

- Exposure sample analysis will help determine if micronized version of ABC-123 has higher bioavailability and may be the reason for the liver toxicity.

FIGURE 11.9   Conclusions

Timeline
- Histopathology has just started
  - Paying extra fee for expedited reading
  - Unaudited results and representative photos will be provided as soon as the liver slides are read

- Exposure analysis will be completed in two weeks

- Draft report will be provided in two months

FIGURE 11.10   Study timeline

STUDY AUDIT TEMPLATE

Nonclinical Study Audit Report

Study Number

Study Title

Test Article

Study Director

Testing Facility

Sponsor

Date(s) of Study Audit

Individuals Involved

Purpose

Audit Findings

Conclusion

{Study Monitor}   Date
{Title}

Chapter 12

A Study Director's Perspective on Study Monitor–Study Director Interactions

Mark Morse PhD, DABT

Charles River Laboratories, Spencerville, OH

Key Points

- An effective working relationship between you and the Study Director is critical to the success of the study. Establish your expectations and confirm your timelines with the Study Director as early as possible
- While the CRO is willing to use your company-specific report templates, the CRO's protocol templates should be used to avoid confusion or errors on the part of the technical staff
- Although you and your company "own" the study, the Study Director must be able to demonstrate effective command and control over the study; this includes any study phases conducted at Test Sites that you may have contracted with directly

---

ESTABLISHING THE WORKING RELATIONSHIP

As indicated in Chapter 1, establishing a good rapport with the Study Director will be critical to your effectiveness as a Study Monitor. For the purposes of this chapter, Study Monitor is synonymous with the Contracting Scientist terminology used in earlier chapters. It is possible that more than one Study Monitor exists for a study such as when a company's Toxicologist is responsible for the overall design of the study but a dedicated company Study Monitor is assigned to keep track of the progress of the study and physically visit the laboratory during the in-life portion of the study. Although Study Directors are generally trained and encouraged to provide the best customer service that they can, they will find it much easier to interact with people with whom they are genuinely comfortable. The Study Director *expects* to be your advocate as well as your

Nonclinical Study Contracting and Monitoring. http://dx.doi.org/10.1016/B978-0-12-397829-5.00012-0

eyes and ears at the testing facility; the more cordial and collegial your relationship with the Study Director, the easier his/her job will be and the easier *your* job will be. Also, Study Directors who are genuinely comfortable in working with you are much more likely to go the "extra mile" for you than those who can barely tolerate you. You should *never* underestimate the power of being liked. Most Study Directors will automatically assume that you are demanding, as most of us expect that any effective Study Monitor should be at least somewhat demanding; however, we find Study Monitors who are demanding but likeable to be infinitely more bearable than those who are demanding and unlikeable.

In North America, it is most common for the Study Monitor and Study Director to address each other on a first-name basis. If you are not comfortable with this approach, you should make this clear to the Study Director. In practice, many Study Directors will initially err on the side of formality until some signal is received that indicates that a first-name basis is acceptable to the Study Monitor. This is particularly the case with foreign clients, especially certain East Asian and Pacific Rim clients, to whom the Study Director would be expected to maintain a high degree of formality until the Study Director is instructed otherwise.

If you feel that you are having great difficulty in working with the current Study Director, you can request that the Testing Facility's Management name another Study Director, although this option should only be exercised in extreme circumstances. Ideally, both you and the Study Director should feel as if you are part of the same team.

SETTING YOUR EXPECTATIONS

Perhaps one of the most important initial steps in building the Study Monitor–Study Director relationship is the setting of expectations. You will have fewer problems with the Study Director if you communicate your expectations *clearly* and *early* in the working relationship. A seasoned Study Director can be expected to bring up the topic of expectations very early in the process; if he/she fails to do so, the Study Monitor should not hesitate to broach the topic. In setting expectations it is best to have a formal discussion rather than an exchange of e-mails. Therefore, telephone conversations and site visits, particularly if the visit occurs prior to study initiation, are generally the best venues for informing the Study Director of the Study Monitor's expectations. If the Study Monitor's company has an explicit document that details Contract Research Organization (CRO) expectations, the Study Monitor should ensure that the Study Director has access to the most recent version of the document prior to the discussion. Typical topics that will be covered in the expectations discussion will include expected timelines, communication logistics, frequency of updates, and any other pertinent study-related areas.

Normally, the expected timelines for the study or program will have already been provided to the CRO's business unit prior to proposal generation and then forwarded to the Study Director after the study was awarded to the CRO. However, it is a good practice to confirm that the Study Director is aware of your timelines, particularly the expected start date and the expected draft report delivery date. Also, if you have an expectation that the report must be finalized by a certain date, be certain that you provide that date to the Study Director. If you expect to submit a draft report to any regulatory agency, inform the Study Director in advance (see *Reporting* later in this chapter). If your timeline should be delayed for any reason, you should inform the Study Director as soon as is practical.

Communication logistics should be established as early as possible in the working relationship. At a minimum, you should expect to provide your e-mail address and business telephone number. If you expect to be contacted via telephone for critical study matters after hours or on weekends, be prepared to supply your home and/or cell phone number as well. If you expect to be unavailable during any critical study phases, it is essential that you provide the Study Director with a back-up contact who can make decisions in your absence. It is also useful to explore various *what if* scenarios with the Study Director for those situations in which you or your back-up cannot be reached for a critical study event in a timely manner. When discussing communication, state any explicit expectations that you may have for the types of study-related information that you would like to be provided. The conscientious Study Director will tend to err on the side of providing you with more information than you may otherwise expect. It is reasonable for you to expect prompt responses to any of your inquiries that are delivered via voice mail or e-mail; it is an industry expectation that CRO Study Directors should respond to all inquiries within 24 hours of receipt.

THE STUDY DIRECTOR'S EXPECTATIONS OF YOU

Chapter 1 pointed out that Study Directors can vary considerably in their perceived strengths or weaknesses. Study Directors also find that Study Monitors differ considerably in their communication and organizational skills, timeliness, training, education, and experience. Experienced Study Directors have encountered both effective and ineffective Study Monitors at all levels of training and experience. From the Study Director's point of view, the most effective Study Monitors tend to excel in organization, communication, responsiveness, and knowledge of CRO procedures and Good Laboratory Practices (GLPs). Listed below are four important points that most Study Directors would like you to consider:

1. *Accessibility, responsiveness, and timeliness matter as much to the Study Director as to they do to you.* For an effective working relationship, accessibility and responsiveness must be a two-way street. Just as you expect to be able to reach the Study Director without undue difficulty and have the Study Director respond to you in a timely manner, the Study Director will

require the same of you. In cases of adverse or unforeseen study events, it is critical that the Study Director be able to contact you in a timely manner and reach a consensus on any necessary study changes. Also, Study Directors will greatly appreciate your *timely* review of draft protocols and amendments, as we cannot finalize them without your approval. If an amendment is urgently needed, the number of cycles of revision and review must be kept to a minimum.

2. *Understand the Study Director's role.* In the GLP environment, the Study Director is the single point of control for all study matters, and to maintain GLP compliance, the Study Director must therefore be able to demonstrate effective command and control of the study. This is particularly important for both the Study Director and Study Monitor to remember in dealing with Sponsor-designated Test Sites. Although you may have subcontracted third party contributing scientists or arranged for the participation of your own company's scientists for certain study phases, the Study Director must be kept aware of all of your interactions with such contributing scientists and it is important that any instructions for study-related matters emanate from the Study Director rather than from you. It will be impossible for a Study Director to claim control of a study phase conducted at a Test Site, if the entire study phase is conducted without any input from or contact with the Study Director. Every good Study Director takes GLP compliance very seriously; you should as well, as it is in the best interests of your study and your company and critical to the overall success of the study.

3. *Avoid significant last-minute study changes.* Every Study Director knows that urgent protocol changes during the conduct of a study are occasionally necessary due to adverse test article effects or unexpected circumstances, and we strive to be prepared for such occurrences. What can drive a Study Director to great distraction are massive, last-minute changes prior to study start that are implemented solely because you and your company were not certain what endpoints or procedures were truly desired from the beginning or because the study was rushed into production. The danger of such last-minute changes is not merely the additional administrative work on the Study Director's end, but also that the plans and schedule of the technical staff must be rapidly altered to accommodate your changes, which increases the possibility of study errors. Technical resources are initially allocated based on the study design that the CRO bid on and that you initially proposed or accepted; changes such as increasing the time, complexity or number of study procedures, increasing the number of study animals, or changing the species may not be easy or even possible to accommodate within the context of the original study initiation timeline. Also, such last-minute changes may generate additional protocol amendments, as the shortened review time may not allow the Study Director and CRO personnel sufficient time to examine thoroughly the proposed changes prior to protocol finalization.

4. *We respect your time – please respect ours.* As pointed out in Chapter 1, nearly every CRO Study Director will have more than one study running at a time and other customers to serve. Any dedicated Study Director will attempt to provide you with the best service that he/she can, as customer satisfaction and repeat business are critical to the Study Director's and the CRO's reputation and financial survival. However, we will not be able to dedicate every minute of every hour of every day to you and your study, whether you are presently on site monitoring your study or off site. If you really must observe that toxicokinetic sample being drawn at 2 a.m., we will try to accommodate you, but we would greatly appreciate having some advance notice. If you insist on watching all 1200 of your rats being dosed on the first day, be aware that the Study Director may have to designate someone else to accompany you in the animal room for at least part of that time.

PRE-PROTOCOL COMMUNICATIONS

Prior to protocol generation and submission of the study protocol (or other CRO-specific forms) to the CRO's Institutional Animal Care and Use Committee (IACUC), the Study Director will normally request information from the Study Monitor. The Study Director should already have a study design furnished by the Study Monitor or the CRO's business group. Items that are commonly requested by the Study Director may include, but are not limited to: identification of test and control articles (if not supplied previously); the therapeutic indication of the test article, (or purpose); the test article's mechanism of action; certificate(s) of analysis (or equivalent); stability information on the bulk test article(s); material safety data sheets (or equivalent safety and handling instructions); a summary of effects observed in any previous studies conducted with the test article(s); and the results of any literature searches conducted on the test article(s).

You may have already generated your own estimate for the quantity of test article(s) that the study requires. However, it is best to confirm that estimate with the CRO laboratory personnel. The Study Director may generate such an estimate although, frequently, the estimate will be provided by the CRO's formulations/pharmacy group. Bear in mind that the person preparing the estimate will attempt to account for unavoidable waste, dead volumes in syringes, intravenous catheters, etc., and normally, a certain excess (e.g. 20–25%) will be assumed. If there is a large discrepancy between your test article estimate and the CRO's, the Study Director can supply the Study Monitor with the actual calculations so that you can understand the assumptions that the laboratory uses in preparing test article estimates. The CRO can often reduce the estimate when test article supplies are known to be limited in advance, but the Study Monitor should be aware that the laboratory personnel are normally the best judge of how much test article will be required to conduct the study.

If the test article for the study is being shipped to the laboratory from another country, be certain to account for possible delays in the delivery schedule, as delays in clearing customs (either in the country of origin or the destination country) are not uncommon in this era of heightened security concerns. Many CROs as well as Sponsors employ customs brokers specifically for such cases. Also, the test article may be shipped to the CRO under a Food and Drug Administration (FDA) hold, so that the CRO laboratory can only store the material until written notification from the FDA allowing use of the material is received. Inform the Study Director in advance if your test article will be shipped from a foreign country and if you have already engaged the services of a customs broker.

PROTOCOL DEVELOPMENT

Normally, the Study Director will write the initial protocol draft, using a study design provided by the contracting company. Certain areas of the draft protocol will be highlighted for the Study Monitor/Contracting Scientist to address, such as dose levels, dose volumes, dose formulation concentrations, the names and contact information of contracting company-selected principal investigators. When revising the draft protocol, it is important that the Study Monitor make only those changes that are absolutely necessary to ensure that the contracting company's required procedures are clearly stated. The protocol is, first and foremost, the study instruction manual for the technical staff. The protocols of most CROs contain a good deal of procedural detail taken from stock templates that the technical personnel are accustomed to seeing in their protocols. Rewriting such standard phrasing solely for the sake of literary flow will not add much, and may actually result in error or confusion on the part of the technical staff; a good Study Director will be certain to remind the Study Monitor of this possibility. Any revisions or additions the Study Monitor makes should be as clear or concise as possible, and the Study Monitor should expect that the Study Director may make further revisions to the Study Monitor's changes in order to make them more readily understandable to the technical staff. In addition, the technical personnel are used to seeing the protocol instructions listed in a specific order; changes in the order of the protocol instructions can lead to serious technical errors, and you can expect the Study Director to resist such changes. For these reasons, contracting companies are *strongly* advised not to use their own protocol templates but rather to use the CRO's existing protocol templates to construct the draft protocol, and to keep any substantial revisions to a minimum. As the Sponsor representative, you will have to approve formally the protocol, and the date of your approval must be included in the protocol; most laboratories find it simpler to have the Sponsor representative sign the protocol, as the Sponsor representative's signature and date is a clear indication of Sponsor approval. If you elect not to sign the protocol, you should provide the Study Director with your written approval (an e-mail maintained in the study records will suffice), and the date of your approval will be included in the protocol.

As indicated in Chapter 2, intentional revisions of the study protocol are normally referred to as protocol amendments. As a best (and safest) practice, protocol amendments should be *prospective* rather than *retrospective*, i.e. the amendment should be signed and distributed to study personnel in advance of the intended change. Although deviations are normally considered to be mistakes or unintentional departures from protocol-required procedures, simple one-time changes can be intentionally made by the Study Director and documented as a protocol deviation. If the intended change is expected to recur, the deviation should be followed by a protocol amendment (if not completely replaced by the amendment) in order to implement the recurring change. As with the protocol, the Sponsor representative must approve the amendment, although there is no regulatory requirement for the Sponsor representative actually to sign the amendment; indeed, securing the Sponsor representative's signature in advance can unnecessarily delay an urgent amendment. The Study Monitor and Study Director should both attempt to avoid large, substantial protocol changes at the last minute before study initiation, unless such changes are absolutely necessary. Last-minute changes can be very unsettling to the technical personnel, and entail a greater risk of initial study errors.

THE PRE-STUDY MEETING

Prior to study initiation, the Study Director will normally hold a pre-study meeting to discuss the details of the study with the study personnel. Attendees generally include the Study Director (who will chair the meeting), representatives of each relevant department, and principal investigators or contributing scientists from any remote test sites that may be involved. Although the Study Monitor's participation in such meetings is not mandatory, it is a good opportunity for the Study Monitor to make his/her expectations known to the technical staff in advance. Also, by attending the pre-study meeting, the Study Monitor can quickly get an idea of how the Study Director may approach the conduct of the study by the manner in which he/she runs the meeting. Normally, the Study Monitor will participate in the pre-study meeting by teleconference. The technical representatives tend to appreciate any information on the intended therapeutic indication, purpose, and class of the test article, as this helps them put the objectives of the study into the appropriate context. If the Study Director does not provide such information, the Study Monitor should be prepared to do so. During the meeting, the Study Monitor should make certain that the Study Director emphasizes any protocol-specified procedures that are not routine for the laboratory, and should carefully listen to the Study Director's answers to any specific questions and be prepared to add any clarification or further explanation. Additional pre-study meeting topics often include relevant safety precautions for test article use and handling, and prior knowledge of or the anticipation of any possible pharmacologic or toxicologic effects. For a multisite study, the schedule of sample/specimen shipment should be reviewed as well. The Study

Monitor should not be surprised or overly concerned if the pre-study meeting identifies a few protocol items that may require changes that result in a protocol amendment. Although the Study Director will typically distribute the draft protocol to the technical staff for comment prior to signature, it is natural that the technical staff's focus on the actual details of the study protocol will intensify just prior to study initiation.

VISITING THE LABORATORY

Inform the Study Director of your visit as far in advance as possible, so that the necessary arrangements can be made. In practice, the Study Director and CRO are in no position to refuse your visit, as your organization is paying for the study. Be aware that the Study Director could have more than one Sponsor visiting on a given day, and therefore may not be able to spend the entire day with you. The Study Director or the CRO's business group should be able to provide you with driving instructions and suggestions on airports, airlines, and hotel accommodations. If you require transportation to be provided, let the CRO know this in advance as the standard expectation is that the Study Monitor will secure his/her own transportation. It is standard practice for the CRO to provide you with lunch on each day of your visit. However, you should inform the Study Director of any dietary restrictions that you may have.

If you have not visited the laboratory previously, request a tour. Bear in mind that certain areas will be off limits to you, particularly other Sponsors' study rooms or areas where assuring client confidentiality is difficult, for example, bulk test article storage areas as the test articles may be stored in containers that clearly indicate the Sponsor company's name. You normally will not be permitted to enter the Study Director's office, as there may be study-related materials for other Sponsors on the Study Director's desk. You should expect to be escorted at all times while in the facility. The Study Director should inform you of any CRO policies that may prevent you from being admitted to the vivarium. For example, if you expect to tour the vivarium of a facility that houses nonhuman primates, expect the Study Director to request a recent tuberculin skin test.

Be certain to inform the Study Director of all procedures that you expect to see during your visit and all of the personnel that you would like to meet. Also, coordinate your expected time of arrival with the Study Director. If it is your first visit to the CRO, the Study Director will not be surprised if you arrive later than planned. However, the typical Study Director will be taken unawares if you should arrive an hour early. If you had expected to view all study-related procedures during your visit, you should consider a contingency plan that will allow the study personnel to proceed if you should arrive at the laboratory far later than planned; the Study Director will be reluctant to delay excessively daily study procedures, as the technical staff will have other scheduled duties besides your study. If you are visiting to monitor a study, most Study Directors will expect that you may want to see data books, Standard Operating Procedures

(SOPs), and training records of relevant personnel, and these materials should be available for your review.

When viewing technical procedures at the CRO, bear in mind that there is generally more than one way to perform a procedure properly, and the specific manner that you have observed the procedure performed in the past elsewhere may not be the procedure of choice at the current CRO. If you feel that any observed procedure is being conducted improperly, you should bring this to the attention of the Study Director immediately. Although experienced technical personnel should be used to conducting procedures in the presence of Study Directors, Study Monitors, Quality Assurance personnel, or regulatory inspectors, it is not uncommon for technicians to be a bit nervous in front of strangers, particularly at the start of a study. However, displaying a friendly demeanor will help to allay any nervousness on the part of the technical staff.

STUDY UPDATES

Most CRO Study Directors will expect to provide you with regular study updates weekly. In most cases, this would consist of summary tables of study parameters such as survival, clinical signs, body weights, and food consumption with a brief textual description. For an acute or subchronic study, you should be notified of any deaths on the day that they occur and the Study Director should provide a global assessment of the presumed cause of death (e.g. dosing error, accidental injury, suspected test article toxicity). In the last few months of a 2-year carcinogenicity study, it becomes unreasonable from the Study Director's perspective to inform you of every single death on the day of occurrence, as animals will succumb to complications arising from senescence on a daily basis; in such cases, mortality updates may be provided weekly or every other week. For shorter-term studies in which severe toxicity or other ongoing difficulties occur, daily updates may be warranted. It is standard practice for the Study Director or designee to send complete summary tables for all parameters evaluated thus far (including standard clinical pathology evaluations) at the end of the in-life phase.

STUDY ISSUES AND THE UNEXPECTED

Most study issues that arise during the in-life portion of the study tend to fall into three major categories: (1) test article issues; (2) unexpected or severe toxicity requiring prompt study changes; and (3) study errors that may have an impact on the study.

Test Article Issues

One test article issue that is commonly encountered is difficulty in the preparation of dosing formulations. Under normal circumstances, the contracting company will provide the CRO with explicit test article formulation instructions.

However, such instructions may be based on the preparation of small pilot formulations that may not be easily scaled up to actual study use with the equipment that the CRO commonly uses. In addition, the original instructions may have been developed with a previous batch or lot of the test article. Even relatively small variations in the concentrations of impurities can drastically alter the solubility or other characteristics of the test article, requiring procedural changes to produce useable solutions or suspensions. This can also be a problem when different batches of test article are supplied to the CRO during the course of long-term studies (i.e. chronic or carcinogenicity studies).

In addition, test article supply problems can also occur. One nightmare scenario for both the Study Director and the Study Monitor would be to run out of test article before the study is completed. A good CRO formulations/pharmacy unit should accurately estimate the quantity of test article needed for the study and carefully monitor the usage of test article during the course of the study to ensure that there will be sufficient test article to complete the study. However, there are occasions in which the assumptions used to estimate the test article required for the study simply do not mirror actual study conditions, often due to the peculiar characteristics of a given test article. In addition, when the contracting company arranges for periodic shipments of test article during the course of a long-term study, any delay in the expected delivery of the shipment could result in a shortfall in test article if the shipment is not planned well in advance of the expected need for resupply.

Unexpected or Severe Toxicity

The Study Director should contact you (or attempt to contact you) immediately when unexpected or severe toxicity occurs during the conduct of the study. Protocol changes may be warranted, depending on the nature and the severity of the findings. Please note that any significant changes to the study protocol related to animal care and use will have to be reviewed and approved by the CRO's IACUC before such changes can be implemented by protocol amendment. Therefore, the Study Monitor should be mindful that such changes may not be able to be implemented immediately. Examples of significant changes may include, but are not necessarily limited to: changes in study objectives; changes in methods of euthanasia; additional procedures or changes in procedures that may increase the potential for pain or distress; changes in surgical or other invasive procedures; changes in dose levels; changes in sampling frequencies; sampling volumes; restraint times; and changes in the study duration.

Study Errors

In general, you should expect to be informed of any major study error in a timely manner, most often via telephone. If you are not available for a telephone conversation, then the Study Director will likely contact you via e-mail. When informed of a study error, you should expect the Study Director to give you a clear, concise description of what actually happened, his/her interpretation of the impact of the error on the study, and any suggested corrective actions to ensure that there is no recurrence of the error. Bear in mind that any possible remediation strategies such as increases in study duration, addition of another study group, discounts for future studies, or study repeats at the CRO's costs are matters that the Study Director will have to first discuss with Testing Facility Management, so the Study Director may not have a specific remediation plan to offer the Study Monitor (if warranted) at notification of the study error. When the incident involves human error (as is frequently the case), be aware that the Study Director will generally not be privy to the details of any employee action(s) by the CRO's human resources department, nor would the Study Monitor be informed of such details.

REPORTING

In many cases, the Study Director will convene a pre-report meeting with the study pathologist and any contributing scientist/principal investigators as well as the Study Monitor, if requested. The purpose of such a meeting is for the study team to reach a consensus on significant findings, the interrelationship of any findings, and any risk assessment benchmarks such as the NOAEL (no-observed-adverse-effect level), NOEL (no-observed-effect level), MTD (maximum tolerated dose), etc. Although the Study Director is ultimately responsible for the interpretation, analysis, documentation, and reporting of results for the study under GLPs, it is recognized that the Study Director will, most likely, not have expertise in all of the scientific disciplines that the study may encompass. Therefore, the expert opinions of any principal investigators or contributing scientists will be welcomed by the Study Director in crafting the overall study interpretation and conclusions. It is at this stage that you would have the best opportunity to provide any input and insight. Once the Study Director has crystallized his/her interpretation and the draft report is written, the Study Director will be much less open to suggestions on the interpretation of the study and the study conclusions.

If you have a company-specific report template that you expect the CRO to use, provide the template as far in advance of the draft report submission date as possible, and include instructions and/or examples. While CROs are reluctant to use the Sponsor's protocol templates, the use of Sponsor's report templates for reporting is fairly common, especially for long-established pharmaceutical companies. It will be much more work for a CRO to convert a draft report that was already issued in its own stock template to a Sponsor-specific template after the fact than to generate the report using the Sponsor's template from the beginning.

Inform the Study Director far in advance of your expected timeline for finalization of the report, and bear in mind that most CROs generally cannot finalize a large GLP study in a day or two. Be aware that it can take a few weeks to incorporate finalized sub-reports from Test Sites, to perform final Quality Control (QC) and Quality Assurance (QA) reviews, and to prepare study materials for archiving. Normally, the Study Director or other CRO personnel will inform you of the typical time required for finalization long before the desired finalization date. The Study Director and Study Monitor should work together to ensure that any study sub-reports are finalized well in advance of the main toxicology report, as this keeps the process moving. Remember that the Study Director is the person who must ultimately authorize the finalization of any sub-report, even if your company is conducting a specific study phase or even if you may have directly subcontracted a third party laboratory to conduct a specific study phase. Also, the Study Director will require documentation (e.g. a letter or e-mail) from your Test Site that the data for a specific study phase have been archived at the time of finalization of the sub-report for that study phase.

The matter of sub-report finalization is particularly critical in the case of the pathology report. The FDA considers the final signed pathology report to constitute the raw data of the pathology phase[1]; therefore, in the eyes of the FDA, a draft pathology report is meaningless, as the pathologist's findings could change due to the subjective nature of microscopic evaluation. If you plan to submit a draft toxicology report, make certain that the pathology report is finalized, as the FDA has rejected submissions with unsigned pathology reports for the reason above. In addition, CROs have been issued 483 findings for the submission or use of unsigned pathology reports; consequently, your Study Director should be keenly aware of the need for a signed pathology report for any regulatory submission. Because of this special emphasis on the pathology report, it is always advisable to ensure that the pathology report is finalized well in advance of the main report. To finalize the main report on the same day as the pathology report would be a clear indication to any regulatory reviewer that the Study Director *must* have relied upon a draft of the pathology report to prepare the main report and, therefore, the Study Director did not have access to the raw data of the pathology phase (i.e. the signed pathology report).

CONCLUSIONS

Establishing a good working relationship with the Study Director is critical to the success of any study. The Study Director is there to help you and keep you informed about the progress of the study. By setting expectations early and opening clear lines of communication, you will increase the chances that not only will the Study Director deliver what you need but you will also be kept informed about the progress of the study.

---

1. 21 CFR 58: Good Laboratory Practice Regulations; Final Rule. *Federal Register* 52 (172): 33768-33782, 1987.

Chapter 13

Draft Report

**William F. Salminen PhD, DABT, PMP*, Joe M. Fowler BS, RQAP-GLP†
and James Greenhaw BS, LAT†**

**PAREXEL International, Sarasota, FL, †National Center for Toxicological Research, FDA,
Jefferson, AR*

> **Key Points**
> - It is important to set expectations for draft report quality, format, and timeliness with the laboratory and Study Director before the study starts
> - Following a systematic process for reviewing the draft report and comparing it to the summary and raw data helps ensure that the draft report accurately represents the conduct of the study and the data generated
> - A clear method should be used to track changes to the draft report and communicate them in a legible manner to the laboratory

The previous chapters dealt with designing and running a clean study that passes regulatory scrutiny. These next two chapters deal with the most critical aspect from a product approval standpoint – reporting the results to support the safety, efficacy, or some other aspect of the test article or treatment. The study report is critical since it is typically the first contact a regulatory reviewer has with the study. Unless the study protocol was submitted for regulatory review, which is rare for most human drug nonclinical studies, the reviewer will have to understand how the study was conducted and also determine the results of the study. A clearly written study report sets up a solid foundation with the reviewer since it is a reflection of how well the study was conducted. A poorly written report raises red flags and makes the reviewer dig deeper into the report since they need to make sure the study is not terminally flawed.

This chapter deals with setting expectations for the draft report and conducting the review. These are the first steps towards achieving a superior report that clearly outlines the study design, reflects the quality of the study, and reports all of the necessary results required to assess safety or efficacy.

Nonclinical Study Contracting and Monitoring. http://dx.doi.org/10.1016/B978-0-12-397829-5.00013-2
2013, Published by Elsevier Inc.

SETTING EXPECTATIONS

Expectations for the draft report should be set as early as possible, ideally during the Project Proposal stage (see Chapter 6). Expectations should be set for at least timeliness, quality, content and format.

Timeliness

The laboratory needs to understand your timelines and requirements for receiving a draft report. Good Laboratory Practice (GLP) studies are tightly controlled and follow regimented protocols and timelines. This greatly facilitates meeting specific milestones and deadlines. Therefore, it is not unreasonable to require the receipt of a draft report at a defined time after the in-life portion of the study has been completed. Setting this expectation during the project proposal stage helps the laboratory price the study accurately and also ensures that they plan accordingly for meeting the deadline. It is helpful to put timelines within the contract and/or protocol so that the laboratory clearly understands their obligations. This also helps you reassure your project development teams and management that they will receive results at the expected time.

When setting the deadline for receiving the draft report, it is best to set this based on the number of days after the last in-life study function has been completed (e.g. last necropsy). The last in-life study function is often fixed providing a solid basis for your timeline. If it is based on a study function that is not tied to a hard date in the protocol, the laboratory can still meet their obligations without meeting your timelines. For example, if the laboratory agrees to provide a draft report three months after the completion of the histopathology slide readings, several months could elapse between the end of the in-life phase and the start of the histopathology readings, greatly delaying your timeline.

The laboratory will provide you with information on how soon after the last in-life function they can realistically provide a draft report. Some laboratories will charge an additional premium for a faster turn around. In general, the shorter the in-life portion of the study, the quicker you can get a draft report since there will be fewer data to compile. Most laboratories will provide you with their typical reporting timelines. The key is to get a fixed timeline and have the laboratory commit to the timeline, which is where inclusion into the contract helps.

Quality

Setting report quality expectations is critical to getting a solid draft report that accurately reflects the study. It is important always to remember that for-profit contract laboratories are under pressure to conduct the maximum number of studies using the least amount of resources as possible. The counter forces to this profit maximization are maintaining a laboratory's reputation and complying with regulatory requirements. Depending on how the laboratory wants to balance these different criteria, the quality of the draft report can vary greatly from laboratory to laboratory. Setting expectations for the quality up front will allow you to get a high quality draft report even from a laboratory that typically pushes out poor draft reports simply to meet timelines.

When setting your quality expectations, it is helpful to provide the laboratory with a copy of a well written draft report that meets your expectations. The example draft report can be redacted so as not to reveal proprietary information.

The quality of the data review and scientific interpretation also need to be established. At a minimum, all of the data in the draft report should go through a full quality control (QC) review. Ideally, the draft report should receive a thorough quality assurance (QA) review, using the same scope and criteria as for the final report. This helps ensure that you are not wasting your time reviewing inaccurate data.

The scientific interpretation of the study results varies widely depending on the laboratory and the Study Director. Some Study Directors excel at scientific interpretation; whereas, others might run a great in-life study and have stellar data, but their interpretation of the data is lacking. Although it is difficult, if not impossible, to dictate the caliber of the scientific interpretation, you can at least make sure that the data are accurately reported and summarized so that you can easily review the report and assist the Study Director with making more solid scientific interpretations, if needed.

Many laboratories have a separate Report Writing Group that compiles the report. It is important to remember that many of the report writers are not scientists and they are simply compiling the different portions of the study data into a report. Depending on the involvement and aptitude of the Study Director, they may or may not revise the report coming out of the Report Writing Group before it comes to you. It is critical that you insist that the Study Director reviews the draft report and revises the report, as needed, so that it makes sense scientifically.

Format

Many laboratories have a default study report format they will use unless you request otherwise. Since the laboratory is intimately familiar with their reporting format, it is often easiest to use their format. As mentioned previously, the people in the Report Writing Group of most laboratories are not scientists and often they have not even worked in the laboratory. They are often very good at compiling all the pieces of a report using the laboratory's default format. If you ask them to use a different format (e.g. a company-specified report template), be aware that it may require many drafts before they understand your requirements and how to compile your report.

If you require a specific report format, it is important to discuss this with the laboratory during the laboratory selection or contract phases so the laboratory clearly understands your requirements and they can make any necessary cost adjustments. Certain regulatory requirements may also require a specific reporting format or inclusion of certain pieces of information that are not typically included in a study report. For example, certain Japanese regulatory agencies require a very rigid reporting format and the US FDA Center for Veterinary Medicine requires the inclusion of all hand-captured raw data as an appendix to the study report.

The final formatting issue that needs to be considered is whether you want a paper or electronic formatted report. Both Microsoft Word and Adobe Acrobat provide very useful editing and commenting tools for electronic documents. Dual computer monitors also make reviewing and manipulating multiple pages and sections of the report much easier than in the past. However, it is a personal preference on the best format for reviewing a report. Sometimes it is helpful to have multiple appendices laid out for quick reference as opposed to scrolling to find them on a computer monitor. Regardless of the approach, it is important to let the laboratory know if you have requirements that differ from their standard operating procedures.

RECEIVING THE DRAFT REPORT

If you decide to receive a paper copy or an electronic copy on physical media (e.g. CD ROM), the laboratory will typically send the documents via a courier that tracks the physical delivery of the documents. If you decide to receive an electronic copy and the laboratory is transferring the documents over the Internet, it is important to understand the potential security risks. Many laboratories and companies have methods for securely transmitting electronic documents and it is best to use those. Attaching the draft report in an e-mail is not very secure unless you have special security capabilities. It is best to check with your information technology department before agreeing to let the laboratory transmit a draft report to you via e-mail. In addition, size restrictions may preclude sending the report via e-mail.

Once you physically receive the draft report, you will need to keep track of it and any changes and comments you make. Some companies have very sophisticated systems for tracking all versions of the report, including uploading an electronic version of the report into a database. Other companies leave it up to the individual scientist to decide how to maintain the draft report. Regardless of the method, it is important to make sure you can track any changes and comments so that they can be transmitted to the laboratory in a clear and legible manner.

Keeping Track of Changes and Comments

There are various ways you can track your changes and comments and then communicate them to the laboratory. The following are just a few examples that work well. If you are reviewing a paper report, you can physically mark up the paper copy. Although it is tempting to send this marked up copy to the laboratory in order to save time, in the long run, it is likely to cause confusion and delays due to difficulty in understanding either a person's handwriting or the comments. Space in a draft report is limited (draft reports are rarely double spaced) and it is often difficult to write down all your changes and comments legibly.

A better way of communicating your changes is by using a separate typed document to communicate your changes. This is similar to the way a scientist responds to a reviewer's comments in a peer-reviewed publication. Basically, you list the page, paragraph, section, or other identifying information so that the Study Director understands what portion of the report you are referring to. You then write down your changes and comments for that part of the report. This regimented approach clearly communicates your changes to the Study Director and they can in turn provide responses to each of your comments using the same format.

If you are using an electronic copy of the report, there are several great tools not only to help you communicate your comments but track the changes made by both yourself and the laboratory in different versions of the draft report. In Microsoft Word, the "Track Changes" feature under the "Tools" menu provides tools for easily tracking changes you make to the document (e.g. insertions, deletions, and revisions) along with any comments you make. Assuming the laboratory provided you with a Word version of at least the main body of the draft report, using track changes greatly facilitates revising the draft report and making sure that the laboratory made all the changes they agreed to make.

If you have an Adobe PDF version of the draft report, Adobe Acrobat provides some good editing and commenting tools. It is important to have the laboratory compile the PDF document in a format that is compatible with the editing tools. For example, simply scanning the paper copy of the draft report into a simple PDF will treat each page as an image and the text cannot be recognized by the editing tools.

Even if you receive a Word version of the draft report, many of the attachments and appendices may be sent in PDF format. If this is the case, using a combination of track changes in Word and the Acrobat editing/commenting tools for the remainder of the report works very well.

Once you have received the draft report, it is time to thoroughly review it to ensure that the report accurately reflects how the study was conducted and that all the results and conclusions match the data. Although some regulatory bodies may not conduct a thorough review of the report (e.g. making sure the summary tables accurately reflect the raw data), it is best to assume that they will review the study with a fine-tooth comb and that any data inconsistencies they find are a reflection of the quality of the study. Once a reviewer finds an error, they are more likely to dig deeper into the study, which increases the chances of them finding more errors and potentially disqualifying the study. Therefore, it is highly recommended to conduct a thorough review of all versions of the draft report using the regimented process that follows so that errors are caught early and do not propagate to the final report.

Many reports will contain contributing scientist reports for various functions that may not have been conducted at the main laboratory or need to be stand alone sub-reports (e.g. clinical pathology, pathology, formulation analysis). You should review these contributing scientist reports in the same manner as you review the entire report to ensure that each one is in order. It is ideal if you can review and finalize each contributing scientist report before the main report is drafted since the data and conclusions in the main report will be based on finalized contributing scientist reports. However, if the laboratory sends draft contributing scientist reports along with the draft main report, you will need to make sure that any changes made to contributing scientist reports are accurately reflected in future drafts of the main report.

Quality Assurance Unit Review

This chapter will not deal with the GLP Quality Assurance Unit (QAU) review of the draft report. However, it is important to understand what the GLP QAU review is and is not. Before a draft report is released from a laboratory, it is often reviewed (audited) by the laboratory's QAU. It is also possible that your company may have an internal QAU that will review the report in parallel with the Contracting Scientist's review.

The QAU review is very helpful since it helps find data inconsistency errors between the main body of the report, appendices/attachments, and the raw data. However, it is important to remember that every QAU has different philosophies and criteria for reviewing draft reports. Some may review a limited amount of the data during the draft report stage; whereas, others conduct an extensive review, similar to the one they would conduct on the final GLP study report. Also, some reports may be released as unaudited (e.g. interim study reports for a chronic toxicity study) and these may not receive a QAU review.

It is important to understand the type of QAU review that will be conducted so you can adjust your review accordingly. Also, don't be fooled if a QAU states that they review 100% of the data. Every time the authors have probed a QAU that made that statement, it ends up that they do a thorough review but they never review 100% of the data. Reviewing 100% of the data would entail countless hours of data review and recalculation. For example, QAU reviewers rarely recalculate or retabulate data derived from validated computer systems. The QAU audit usually starts with a 10% audit and may be expanded if accuracy issues arise.

As with the Report Writing Group, QAU reviewers sometimes do not have scientific training and have not worked in the laboratory on the specific types of studies they are reviewing. They are trained to find inconsistencies between the raw data, protocol, Standard Operating Procedures (SOPs), and the report;

however, they may not understand all the technical aspects of the study. This is where you need to conduct a thorough scientific-based review of the report, which includes a quality control review.

One last comment about the QAU review is that you need to understand the type of working relationship the Study Director has with the QAU. Some laboratories maintain a collegial working relationship; whereas, others maintain the QAU as a black box and very adversarial. There are some aspects of the GLPs that are black and white and not subject to interpretation. However, the authors have seen many QAUs get off track and focus on minutia that do not impact GLP compliance or the quality of the study. If the Study Director has a poor working relationship with the QAU, this can lead to a lot of hair pulling by both the Study Director and Contracting Scientist, especially if changes are being requested that go against the QAU's interpretation.

Scientific and Quality Control Review

Everyone has a different method for reviewing reports and there is no one right way. Some methods may be better for certain types of studies; however, the process that follows maximizes your chances for finding both scientific and data quality issues. As mentioned in the QAU section above, it is nearly impossible to conduct a 100% review of the study and raw data, especially given today's tight project development timelines. The following process focuses on reviewing a portion of the data in detail to ensure that the study was conducted correctly, the QAU review was robust, and that all of the conclusions make sense scientifically.

Raw Data versus Summary Data

Given today's extensive use of electronic data capture systems by most commercial laboratories, it is often confusing what constitutes "raw data". This deserves a brief mention at this point since you will likely be reviewing different types of data during the draft report review. Also, some regulatory bodies might be content with summary data; whereas, others might request copies of all the raw data so that they can recreate the study and verify that the report and all data summaries match the raw data.

Raw data are considered the very first place that data are recorded. The two main types of raw data are paper and electronic. Paper data are typically captured on a predefined form where the person making the observations fills out the appropriate places of the form and then initials and dates the form acknowledging they are the person that made the observation. Figure 13.1 is an example of a simple data capture form with hand written original observations and the recorder's initials and date. It is important to understand that the raw data are the first place the data were recorded. For example, referring to Figure 13.1, if a person wrote the animal body weights on their glove and then transcribed them to the form, the original raw data would be the person's glove. The data on the

MANUAL DATA COLLECTION FORM – BODY WEIGHT

Balance Identification Number _35-21_

Study Number: _123-456_   Study Day: _5_

Animal Identification Number	Time	Body Weight (g or kg) (Circle One)	Observation(s)
01	0801	310.4	N
02	0802	384.0	N
03	0804	448.7	N
04	0805	364.5	N
05	0806	309.7	N
06	0808	341.9	N
07	0811	394.3	N
08	0813	330.3	N
11	0814	350.0	N
12	0816	309.3	N
13	0817	266.0	N
14	0820	322.2	N
15	0821	310.5	N
16	0823	293.0	N

N = Appears Normal

Performed By: _JJG_ _01 AUG 2011_ _____ Reviewed By: _KSS 01AUG2011_
            Initials   Date   Initials   Date

Back Entered into Computer System on _02 AUG 2011_ _____ by _JJG_
                                      Date   Initials

Verified By: _____KSS_____ Date: _02AUG-2011_

SOP #1234
Effective: 01-JAN-2011

FIGURE 13.1   Example of hand captured raw data (body weight) from a 5-day rat study

form would then be considered transcribed data and would have to be acknowledged as such. This may seem like a minor distinction, but it is important to make sure the laboratory is recording data correctly.

Electronic data can be confusing since some people think that any data in a computer file are raw data. Only data captured by a fully validated and traceable computer system are considered raw data. For example, referring to Figure 13.1, it is not acceptable to enter the animal body weights into a regular Microsoft

Excel spreadsheet since the data can be subsequently altered without being able to trace who made the change and what the change was. If electronic data are being captured in your study, make sure that they are being captured with a validated system. If not, insist that the laboratory capture the data on paper otherwise you risk having your study being invalid.

Assuming your electronic data were captured in a compliant manner, the raw data are the electronic data, not the printouts of the data. Therefore, if a regulatory body requests copies of your raw data, make sure that they mean printouts or summaries of the electronic data (see Figure 13.2 for an example) and not the electronic files that are often only readable by proprietary software. Most reviewers are content with summaries of the raw data as opposed to the actual electronic file as long as the data were captured with a validated system. Some reviewers might request a generic electronic output of the raw data (e.g. tab or comma delimited file) that can be universally imported into various programs so that they can run their own statistical analyses on the data.

Report Review

Although there are many ways to review a report, the following process ensures that the report makes scientific sense and accurately reflects the raw data. For the sake of simplicity, it will be assumed that you are reviewing a paper copy of the report; however, the same process would be used for an electronic version but the editing and commenting would differ.

STEP 1 – Check That All Major Components of the Report are Present

Ensuring that all major components of the report are present is simple but very important since some laboratories may push draft reports out the door in the effort to meet a contractual deadline. Without a complete report, it is impossible to accurately review the report. While you are conducting this review, you can also make sure that all formatting throughout the document is correct (e.g. pagination, sequential heading numbers, table of contents matches the pages in the report). Table 13.1 lists typical key report components.

It is important to note that these general headings must contain all of the information required under the FDA GLPs as described in the following list (this list was taken from the US FDA GLPs as found at 21 CFR 58.185):

1. Name and address of the facility performing the study and the dates on which the study was initiated and completed.
2. Objectives and procedures stated in the approved protocol, including any changes in the original protocol.
3. Statistical methods employed for analyzing the data.
4. The test and control articles identified by name, chemical abstracts number or code number, strength, purity, and composition or other appropriate characteristics.

ABC Laboratory Study Number 123-456
A 5-Day Oral Gavage Toxicity Study with Compound X in Rats

Individual Body Weight, g – MALE

Group, Animal Number	Study Interval (Day)		
	1	3	5
0 mg/kg/day			
1	291.6	302.5	310.4
2	362.5	370.4	384.0
3	431.5	442.8	448.7
4	345.0	356.4	364.5
100 mg/kg/day			
5	311.1	300.1	309.7
6	355.9	346.4	341.9
7	420.2	395.2	394.3
8	328.6	326.7	330.3
300 mg/kg/day			
9	401.1	320.2	—[1]
10	370.1	300.5	—[1]
11	389.2	352.6	350.0
12	346.2	317.6	309.3
1000 mg/kg/day			
13	307.2	268.1	266.0
14	354.1	327.3	322.2
15	375.9	334.8	310.5
16	346.6	305.4	293.0

[1]Animal removed from study prior to scheduled necropsy

FIGURE 13.2 Electronic summary of individual animal body weights from a 5-day rat study

TABLE 13.1 Typical components of a draft GLP study report

Title Page and General Study Information
Statement of Regulatory Compliance
Key Research Personnel
Table of Contents
Signature Page for GLP Compliance Statement (unsigned at the draft report stage)
Quality Assurance Statement
Signature Page for the Study Director (unsigned at the draft report stage)
Summary
Introduction
Materials and Methods
Results & Discussion
Conclusion
Appendices (see Step 4 for a listing of typical appendices in a toxicology study report)

5. Stability of the test and control articles under the conditions of administration.

6. A description of the methods used.

7. A description of the test system used. Where applicable, the final report shall include the number of animals used, sex, body weight range, source of supply, species, strain and substrain, age, and procedure used for identification.

8. A description of the dosage, dosage regimen, route of administration, and duration.

9. A description of all cirmcumstances that may have affected the quality or integrity of the data.

10. The name of the Study Director, the names of other scientists or professionals, and the names of all supervisory personnel, involved in the study.

11. A description of the transformations, calculations, or operations performed on the data, a summary and analysis of the data, and a statement of the conclusions drawn from the analysis.

12. The signed and dated reports of each of the individual scientists or other professionals involved in the study.

13. The locations where all specimens, raw data, and the final report are to be stored.

14. The statement prepared and signed by the quality assurance unit as described in 58.35(b)(7).
15. The final report shall be signed and dated by the Study Director.

STEP 2 – Compare the Protocol, Amendments, and Deviations to the Study Report

Make sure the draft report contains copies of the signed protocol and any protocol amendments. All protocol deviations should also be listed, typically as an appendix to the report. Review the protocol, amendments, and deviations against the main body of the study report (introduction through conclusions) to ensure that the report accurately reflects what was done during the study and that the protocol was followed. For example, the study report should state how often body weights were measured (e.g. everyday prior to dosing) and this must match the protocol requirements. The only exception is if a deviation has been written that covers a change in the body weight measurement schedule. If a deviation has been written and it has an impact on the study, the report should explain how severe the impact was and the overall affect on the study integrity.

STEP 3 – Summarize the Data in the Main Body of the Report

Read through the abstract and main body of the report and write down all of the relevant details of the study. The following are examples of key items that you want to capture:

- Study identifier (e.g. study number)
- Test article(s) or treatment(s) and relevant identifying information (e.g. batch number)
- Treatment groups
- Species, sex, age, and number of animals per group
- Route and duration of dosing
- Timeline of study events
- Endpoints assessed along with results for each endpoint
- Any unique observations that were not predefined (e.g. adverse events)
- Fate of the animals at the end of the study.

Figure 13.3 is an example of a summary for a relatively simple study. Writing down this information helps you become familiar with the report and also serves as an invaluable reference for ensuring that the report's results match the summary data in the appendices and the raw data.

As you read through the report, make sure to mark up the report with questions, edits, or comments you have. It may be helpful to mark each page that has comments with a flag (e.g. sticky note) so that you do not overlook them when you write up your comments to the laboratory.

Study Number: F09320

Test Article: APAP (Batch #124-A, Expiration 1/1/2015)

Treatment Groups: Control (0.5% methylcellulose), 500, 1000, and 1500 mg/kg body weight.

Route and duration: Oral gavage. Single dose.

Species, sex, age: Rat (Sprague-Dawley), males only, 8 weeks at start of dosing

Number of animals: 5 animals per group

Timeline of events:

SD-7   Physical exam

SD-6   Body weight and randomization to groups

SD-5   Clinical observations (CO) and feed consumption (FC) start

SD-4   CO/FC

SD-3   CO/FC

SD-2   CO/FC

SD-1   CO/FC

SD1   Dose, CO/FC

SD2   CO/FC, necropsy (including blood for clinical pathology)

Endpoints:

- CO – animals in all treated groups lethargic 4h after dose. Lethargy was more severe in higher dose animals.
- FC – decreased feed consumption from SD1 to SD2 @1500 mg/kg
- Clinical pathology – increased serum ALT @1000 and 1500 mg/kg
- Gross pathology – no effects noted
- Histopathology
  - Centrilobular glycogen depletion @1000 (2/5 animals) and 1500 (all) mg/kg
  - Centrilobular necrosis @1500 mg/kg (4/5 animals)

FIGURE 13.3   Summary of key study parameters and results

STEP 4 – Tabulate Summary Data and Data in the Appendices

Ensure that all relevant summary tables and appendices are included in the report and read through each one. Tables 13.2 and 13.3 list summary tables and appendices that are typically included in a toxicology study report.

For the summary tables, write down the key findings. For example, Figure 13.4 shows the liver histopathology findings for a study by group in which rats were administered the test article APAP as a single dose at one of three different dose levels (500, 1000, or 1500 mg/kg body weight) and then euthanized 24 hours after the dose. Two of five animals in the mid-dose group exhibited some degree of centrilobular glycogen depletion. In the high dose group, all animals exhibited some degree of centrilobular glycogen depletion and four of five animals exhibited centrilobular necrosis.

The appendices will vary widely in the amount of data they contain and amount of review that is required. Some appendices may provide relatively simple information that just needs to be verified with the study report and/or protocol (e.g. ensuring that the test article did not expire during the conduct of the study by comparing the expiration dates on the certificate of analysis with the study dates). Other appendices contain data that need to be tabulated to ensure that the report and any summary tables match the data presented in the appendices. Most reports will include the individual animal results for the various endpoints in the appendices (e.g. clinical pathology). The individual animal results will be dealt with later so at this point just review the corresponding summary tables.

TABLE 13.2 Common summary tables in a toxicology study report

Clinical observation findings
Body weight
Body weight change
Feed consumption
Water consumption
Hematology results
Clinical Chemistry results
Urinalysis results
Macroscopic evaluations
Organ weights; organ to brain and body weight ratios
Microscopic evaluations; correlations to macroscopic observations

Depending on the complexity of the study, an appendix may be a sub-report for a portion of the overall study (e.g. dose formulation analysis or histopathology) and may have its own appendices. It is advisable to review each sub-report as a stand-alone study first so that you are reassured that the results and conclusions are accurate. This is a lot of work for a large study, but it prevents errors from propagating through the sub-reports and subsequent references to the sub-reports.

TABLE 13.3 Common appendices in a toxicology study report

Test and control article information (e.g. labels or certificates of analyses)

Dose formulation analysis

Dosing records/exposure calculations

Animal information (e.g. species, strain, sex, dates of birth, animal receipt records [particularly for non-rodents])

Housing information and diagrams (e.g. rack assignment configuration)

Water quality analyses

Feed quality analyses

Animal fate and disposition

Individual clinical observation findings

Individual body weights

Individual body weight change

Individual feed consumption

Individual water consumption

Individual physical examination results and veterinarian interpretive summary

Clinical pathology instrument usage and reference information

Individual hematology values

Individual clinical chemistry values

Individual urinalysis values

Individual organ weights

Individual macroscopic and microscopic pathology and interpretive summary

Statistical report

Historical control values (e.g. clinical pathology)

Protocol and Amendments

Deviations

```
EXP/TEST: F09320                                              PAGE: 1
                                                             DATE: 8/1/11
              NON-NEOPLASTIC MORPHOLOGIES BY ANATOMIC SITE   TIME: 7:51:02
                            PATHOLOGY REPORT
```

MALE CRL-CD/Sprague-Dawley Rat	1 CONTROL 1X, 24H	2 APAP 500 1X, 24H	3 APAP 1000 1X, 24H	4 APAP 1500 1X, 24H
DISPOSITION SUMMARY				
ANIMALS INITIALLY IN STUDY	5	5	5	5
TERMINAL SACRIFICE	5	5	5	5
ANIMALS EXAMINED MICROSCOPICALLY	5	5	5	5
Alimentary System				
Liver	(5)	(5)	(5)	(5)
Depletion Glycogen, Mild, Centrilobular	0	0	1 (20%)	2 (40%)
Depletion Glycogen, Minimal, Centrilobular	0	0	1 (20%)	3 (60%)
Necrosis, Moderate, Centrilobular	0	0	0	4 (80%)

FIGURE 13.4 Summary report of liver histopathology findings from a GLP rat toxicity study

STEP 5 – Verify That the Main Body of the Report Matches the Appendices

Once you have tabulated the information in the summary tables and appendices, you need to compare your summaries with the information listed in the main body of the report. You can do this by comparing the information you compiled in Steps 3 and 4 or you can go back through the main body of the report and verify the information matches the information you compiled in Step 4. If you find inconsistencies, it is important you make a note of the errors so that you can communicate them to the laboratory.

STEP 6 – Track Several Animals Through the Report

Steps 2 to 5 verified that most of the information in the appendices matched the main body of the report. What was left out was a review of the individual animal data. Reviewing and tabulating all of the individual animal data is an onerous task, especially for a large study. If you are including copies of the raw data in the report, the task expands exponentially. This is where it is best to rely on the QAU review to verify that the individual animal data and raw data match the summary tables.

Instead, what you will do in this step is track several animals through all of the study data. If the main body of the report presents results of specific animals (e.g. Animal #123 exhibited convulsions on Study Days 4 to 8), add these animals to your review so that you can verify that the statements are accurate. Use the following process for tracking the animals:

- Select the animals you want to track. Selecting one or two animals per sex per group is usually sufficient. In addition, track any animals that are specifically highlighted in the report.
- For each animal, go through the raw data, individual animal tables, and other relevant appendices and write down specific animal details. For example: vendor, species, sex, date of birth, enrollment physical exam findings, fate of the animals at the end of the study.
- Calculate the doses each animal received. Since some studies require repeated dosing, it is best to select a few days and verify that the animals received the proper dose on those days.
- Tabulate the body weight and feed and water consumption at various times making a note if they increased or decreased over the course of the study.
- Record key clinical observations. Many clinical observations are likely to be listed so it is often best to focus on the ones that have meaning to your study.
- Record clinical pathology results with a focus on the ones that are outside of the normal range.
- Record the gross pathology and organ weight results.
- Record the major histopathology results.
- Record the results of other relevant endpoints.

Once you have tabulated the data for each animal, verify that the data for each animal are consistent with the study report. For example, if you wrote down that Animal #234 exhibited a seizure on Study Day 3, it should be listed in the report and at least one animal should be listed as exhibiting a seizure in the summary tables. If this seizure is not noted anywhere, this is an error that needs to be corrected. Another example is the age of the animals. If the report states that the animals were 7–8 weeks of age during the first dose but the dates of birth are not consistent with that, it is an error that needs to be corrected and possibly written up as a deviation by the laboratory.

STEP 7 – Review the Report for Scientific Merit

If you make it through Step 6 without finding any errors, the laboratory has done a stellar job in writing the draft report from a data integrity standpoint. The final step is to re-review the main body of the report from a scientific standpoint. As you review the report again (you can also do this simultaneously with Step 1 but it is often helpful to conduct a final "big picture" review of the report), determine if it meets your scientific standards. The main question you will be asking is if the overall conclusions are accurate and are they supported by the data. You will also want to determine how much scientific discussion and reference to the scientific literature is required. For example, if you are concluding that a single incidence of a kidney tumor in a high-dose animal was not test article related, you will need to have a lot of scientific references and/or historical control data justifying your conclusion.

Since this is a subjective review, it is left up to the reader to determine the scientific rigor and interpretation required in the report. Regardless, it is important to keep in mind that not all Study Directors have extensive formal scientific training and, even if they do, they might have a different scientific standard than yourself. Therefore, it is important to communicate your expectations to the Study Director and also offer to help them with the scientific write-up and interpretation, if needed.

SENDING COMMENTS TO THE LABORATORY

As mentioned under "Receiving the Draft Report", you need to keep track of your edits and comments as you are reviewing the report. Depending on your preference, you can either send the marked up electronic files (e.g. Microsoft Word with track changes and Adobe PDF with edits and comments) or a separate text document listing your edits and comments. If you are sending electronic files, be aware of potential security risks, especially if attaching documents to e-mails.

Since draft reports often require multiple rounds of review, it is important to label clearly each version with a unique identifier (e.g. the date of issuance or draft number). It is also best to label both the file name and the document itself with the unique identifier so that you can clearly identify all electronic versions of the draft report. This helps you and the laboratory identify the most recent version and any changes that have been made.

Assuming you are sending edits and comments electronically, it is best to insist that the laboratory uses the same method of tracking their changes, at least for Microsoft Word files. This greatly facilitates being able verify what changes were made by the laboratory. Even if you send your comments in a separate text document, it is easiest if the laboratory tracks their changes in the main body of the report using the Microsoft Word track change function so that the changes are clearly listed. Adobe PDFs are trickier since the document itself cannot be modified. In those cases, it is best to have the laboratory recreate any revised PDFs and just state what pages were altered so that you can verify the content.

Once you have sent your edits and comments to the laboratory, they will be reviewed by the laboratory and a decision made about what changes to make. Do not be surprised if the laboratory does not agree to make all of the changes you requested, especially when it comes to interpretation of the results. Under GLPs, the laboratory is an independent body and they have the right to make their own conclusions. They have to correct obvious errors, but they cannot be swayed into writing conclusions they do not agree with (e.g. if an animal experienced ataxia and you think it was a random event not related to the treatment but the Study Director thinks it was treatment related, the Study Director has the right to formulate their own conclusion). These situations are rare but they can happen.

The study contract will often have timelines for various portions of the reporting process (e.g. delivery of the draft report three months after the last animal undergoes necropsy). It is important to be aware of the wording and how this may impact your project timeline. If a laboratory promises to get you a draft report by a certain hard date, there may be no wording on the turnaround time for subsequent drafts or issuance of a final report. It is important either to put hard criteria in the contract or discuss the reporting timeline with the laboratory so you know what the turnaround time will be between multiple draft reports or the last draft report and the final report.

If you find many errors in the draft report and require a lot of revisions, it is best to have the laboratory issue another draft report for your review. This process will continue until you are satisfied that no additional changes are needed or only minor changes are required that you are confident that the laboratory will include in the final report. One area of contention with multiple draft reports is cost. Most laboratories will not charge extra for multiple draft reports if they made errors that need to be corrected. However, if you request changes after the first draft that are not due to outright errors (e.g. your management wants additional scientific discussion in the report and you have already provided comments to the laboratory on the first draft), you might be required to pay extra fees for additional drafts.

SUMMARY

In this chapter, advice was provided on receiving the draft report and keeping track of changes made to the report so that they can be communicated to the laboratory in a clear and concise manner. A rigorous review process was outlined that ensures that the study report accurately reflects the raw and individual animal data. Assuming the review went well and your relationship with the laboratory was collegial, you can rest assured that you have a high quality report that clearly reflects how the study was conducted and the results of the study.

Chapter 14

Final Report, Study Close-Out, and Conclusions

William F. Salminen PhD, DABT, PMP*, James Greenhaw BS, LAT† and Joe M. Fowler BS, RQAP-GLP†

*PAREXEL International, Sarasota, FL, †National Center for Toxicological Research, FDA, Jefferson, AR

> ## Key Points
>
> - It is important to check the final report to ensure that all agreed-upon changes were made
> - The final report should be checked to ensure that it contains all relevant sections and data
> - After issuance, the final report can be changed by issuing a report amendment

The final steps to completing the study you have been working hard to monitor are to have the final report issued and then close out the study. Once you are confident that the latest draft report needs only minor corrections or changes, or maybe even no changes, you can ask the laboratory to make the revisions and finalize the report. One thing to be aware of is that all contributing scientist reports should be finalized (signed and dated) in the latest draft version so that you are confident that these reports will not change. If they have not been finalized, then you might want to ask to see the finalized reports prior to having the main report finalized. At the very least, all of the contributing scientist reports must be finalized before the main report.

FINALIZING THE REPORT

Once you agree to finalize the report, the laboratory will follow their Standard Operating Procedures (SOPs) for finalizing and issuing the report. Many laboratories are eager to finalize a study report since it allows them to close out and archive the study and receive their final payment; however, you should still ensure that the laboratory issues the final report in a timely manner that meets any agreed upon timelines. Once the report is finalized, the laboratory will send you paper and/or electronic copies of the report. Since many regulatory submissions are electronic, it is best if you obtain a PDF version of the final report. Although a simple scanned-in version of the report (i.e. the paper report is simply scanned and converted to a PDF) can be used, it is better if the PDF is created from the original electronic files and includes a hyper-linked table of contents for easy navigation to different sections of the study report. In addition, this type of PDF allows easy searching for keywords. Some pages will have to be images (e.g. the pages with original signatures); however, most will be searchable. Depending on the regulatory authority that the report is being submitted to, electronic signatures may be acceptable; however, it is always safer to also have images of the original hard signatures in case a particular submission does not accept electronic signatures.

Once the final report arrives, you will want to conduct a thorough review of the report to make sure that it is complete. This does not have to be a detailed data review since the draft report review should have caught most of the errors. Instead, you should review the report to ensure that it contains all of the relevant sections, tables and figures, appendices, and contributing scientist reports. You should also make sure that the report is paginated correctly (i.e. all pages are present and in order), the table of contents accurately references the different sections of the report, and section headings are numbered correctly. You should then review the entire report to make sure that every component is present and complete and that any changes from the last draft report you requested have been made. The last thing you want to do is submit a report that is missing a section or part of a contributing scientist report. During the review, you will want to make sure that some of the final components that were not in the draft report have been completed. This includes the signature of the Study Director, study completion date, and dates of Quality Assurance Unit (QAU) audits. Once you are confident that the report is complete, you will want to ensure that the electronic version matches the paper version.

Once you have determined that the final report is complete, you can file it using your company procedures. These can range from simple filing of the paper copy in an archive to sophisticated electronic systems where you upload the final report and it can then be electronically compiled into future regulatory submissions. It is advisable to file the report in a manner that has limited access and can be tracked. This helps ensure that you can locate the report in a timely manner when needed.

As mentioned in previous chapters, the final report is only one piece of the study. There will be paper and electronic raw data, test and control article reserves, and specimens (e.g. histopathology blocks and slides, pharmacokinetic [PK] samples) that need to be archived. The laboratory will archive these items according to their SOPs. Many laboratories will hold these items for a predefined period (e.g. 1 year) and then charge a fee for continued storage. After this period, you will need to make a decision as to whether or not you want to maintain the items at the laboratory or have them transferred to another commercial archive or your archive. You will also need to decide when some samples are no longer usable (e.g. degraded PK samples) and can be discarded. There are various regulatory requirements for the duration of maintaining the raw data and specimens from a study and these vary depending on the regulatory agency and type of application. You should check with your regulatory affairs experts before deciding to discard any raw data or specimens. In addition to the archived items, you are likely to have left over test and control articles. If they will not be used for another study, you will need to decide whether the articles will be held, destroyed, or returned to you.

AMENDING THE FINAL REPORT

As with a final study protocol, once a final report has been signed by the Study Director, it can no longer be modified and reissued. Instead, the final report is modified by a study report amendment. The amendment must clearly identify that part of the final report that is being added to or corrected and the reasons for the correction or addition. The amendment must be signed and dated by the person responsible. Once an amendment has been issued, it should be placed with each copy of the final report in a manner that the reader can clearly determine that an amendment has been made (e.g. at the front of the report). When specific pages of the report are changed, the original pages should not be removed. Instead, the amendment will contain the new pages and will have to be referenced as the original report is read. This can be awkward when reading a final report, but this is the required process under Good Laboratory Practices (GLPs). A similar process must be used for electronic versions of the report.

STUDY CLOSE-OUT

It can be useful to conduct a final audit of the study at the laboratory to ensure that all study-related information and materials are in order and archived correctly. When a report is submitted to a regulatory authority to support a new product approval, the regulatory authority may conduct a targeted inspection of the study, especially if the study is critical in supporting the product approval. The study may also be randomly selected for auditing during a routine laboratory GLP inspection. The regulatory authority will send an inspector to the laboratory to audit the final report and all study-related raw data and specimens. The purpose of the audit it to make sure that the final study report accurately matches the raw data and that the study was actually conducted according to the protocol. Therefore, it is critical that all of the archived information is in order for easy access and review by the inspector.

If you have various studies ongoing at the laboratory, you may be able to piggy back your study close-out visit with a monitoring visit for another study. During the close-out audit, you should review the archived final report and all related study materials to make sure everything has been archived correctly and that everything is present. You can then conduct various levels of audits to determine if the final report matches the raw data. If the study is pivotal in supporting a product approval and is likely to be selected for auditing, you may want to thoroughly audit all the study materials to ensure that there are no issues that will be discovered by the inspector. You can follow a process for auditing the raw data against the final report similar to the process used for reviewing the draft report (see Chapter 13). In addition, you will want to make sure all raw data were captured in compliance with GLPs. For less critical studies that appear to be in order based on the final report and archived materials, you may conduct a more limited spot check of select items. Once you have completed the study close-out, you should write up any findings and conclusions and include them with your study file, as you would for any monitoring visit.

CONCLUSIONS

After reading this book, the novice Contracting Scientist should have a solid understanding of what is involved with designing, contracting, monitoring, reporting, and closing out a nonclinical GLP study. Seasoned Contracting Scientists may have a new understanding of areas that they were not very familiar with. The ultimate goal of this book was to inform the Contracting Scientist about the importance of thoroughly preparing for their study, whether it be from a laboratory selection standpoint to developing a solid protocol and playing an active role in their study once it is running. Simply signing a contract and letting the laboratory run the study on their own terms is likely to result in less than satisfactory results. The advice provided in this book is aimed at providing a foundation for developing, refining, or supplementing existing study contracting and monitoring procedures that may be in place at the Contracting Scientist's company. There is no one right way to implement the advice provided in this book and it is up to the Contracting Scientist and their company to decide what works best. Regardless of the approach, being informed about all the various factors that go into a nonclinical GLP study will help the Contracting Scientist obtain the highest quality studies.

Index

Note: Page numbers followed by "f" and "t" indicate figures and tables respectively

A

AAALAC. *See* Association for the Assessment and Accreditation of Laboratory Animal Care International

AALAS. *See* American Association for Laboratory Animal Science

ACLAM. *See* American College of Laboratory Medicine

ALAT. *See* Assistant Laboratory Animal Technician

Amendment, study protocol, 239
study report, 239

American Association for Laboratory Animal Science (AALAS), 84

American College of Laboratory Medicine (ACLAM), 84

American Veterinary Medical Association (AVMA), 164

Analysis of variance (ANOVA), 166

Animal, Plant, and Health Inspection Service (APHIS), 82

Animal care, 33–34
facilities, 29–30

Animal housing, 63–64

Animal supply facilities, 30

Animal temperament, 61

Animal welfare, 81, 95f–98f
AAALAC accreditation, 83–84
checklist, 94, 99f–101f
day-to-day animal welfare, 84
Guide, salient features, 82
meeting welfare requirements, 81–82
PHS policy and OLAW, 82–83
potential conflicts
analgesics use for animals, 94
animal welfare requirements, 93
GLPs, 93–94
humane endpoints, 94
regulation establishment by USDA, 92–93
salient standards, 84
US Federal Animal Welfare Regulations, 83t
USDA and APHIS, 82

Animal Welfare Act (AWA), 81
regulations established by USDA, 92–93

Animal welfare requirements

and GLP, conflicts between, 93–94
analgesics, use of, 94
humane endpoints, 94
Study Director's role, 93–94

ANOVA. *See* Analysis of variance

APHIS. *See* Animal, Plant, and Health Inspection Service

Assistant Laboratory Animal Technician (ALAT), 84

Association for the Assessment and Accreditation of Laboratory Animal Care International (AAALAC), 83–84

Attending Veterinarian (AV), 85–86

AV. *See* Attending Veterinarian

AVMA. *See* American Veterinary Medical Association

AWA. *See* Animal Welfare Act

B

Backdating, 37

Backup power, 100

Bedding, 88

Bedding materials, 88

Body weight, expected, 157

Business ethics, 144

C

Cageside observations, 162

CDA. *See* Confidential disclosure agreement

CDER. *See* Center for Drug Evaluation and Research

Center for Drug Evaluation and Research (CDER), 40

Certificates of Analyses (CofA), 175, 181–182

Chain-of-custody procedures, 178–179
test article/mixtures, 179

Clinical pathology assessments, 69–70

CofA. *See* Certificates of Analyses

Computer systems, 13, 26

Confidential disclosure agreement (CDA), 109–110, 138
example of, 145–148
MTA, 139
one-way or two-way agreement, 138–139

Confidential information disclosure agreement, 145b–148b
Contaminants, 159
Contract Research Organization (CRO), 2, 234
  in business ethics, 144
  GLP-compliant study, 60, 206
  nonclinical studies
    employee turnover and covering weekend functions, 7
    regulatory compliance and scientific integrity, 6
  selection, 20–21
  Study Monitor–Study Director relationship, 206
Contracting Scientist, 1–2
  CRO, 6
  Study Director, 8
Contracts, 139
  Contracting Scientist, 139
  CRO, 140
  key component, 140–141
  MLA, 142
  partially-completed draft report, 141–142
  protocol and histopathology slides, 141
  sponsoring company, 140
  study price and payment schedule, 140
  test article, 141
Control articles, 23, 155–156
  description of test article, 155
  identity, 155
  reserve sample, 156
  test article disposition, 156
  test article properties/characterization, 155–156
  test article/dose formulation preparation, 156
CRO. See Contract Research Organization

D

Data audit inspection, 40
Data confidentiality, 137–138
  business ethics, 144
  CDA, 138
  confidential information disclosure agreement, 145b–148b
  contracts, 139
    Contracting Scientist, 139
    CRO, 140
    key component, 140–141
    MLA, 142
    partially-completed draft report, 141–142
    protocol and histopathology slides, 141
    sponsoring company, 140
    study price and payment schedule, 140
    test article, 141
  maintaining confidentiality during study, 142–143
    compromise confidentiality, 143
    Study Director and Test Facility Management, 143
  MTA, 139
  one-way or two-way agreement, 138–139
Data management, 193, 195–196
  electronic data files, 196
  nonclinical study audit report, 203
  related information, 196
  study audit, 196
Day-to-day animal welfare, 84
DHHS. See US Department of Health and Human Services
Directed inspections, 40
Dose Formulation Analysis, 174–175
Dosing, 67–68
Draft protocol, 150–151
Draft report, 218
  format
    final formatting issue, 220
    Report Writing group, 219
    requirements, 219–220
  QAU review, 222–223
  quality
    data review, 219
    profit maximization, 218–219
    scientific interpretation, 219
  raw vs. summary data, 223
    electronic data, 224–225, 224f, 226f
    hand captured raw data, 223–224, 224f
  receiving, 220
  report review, 225
    appendices, 233
    FDA GLP, 225
    main body of report, 228, 229f
    protocol requirements, 228
    report components, 225, 227t
    for scientific merit, 234
    summary tables, 230, 231t, 232f
    track several animals, 233–234
  scientific and QC review, 223
  sending comments, 234
    sending edits, 235
    timelines, 235
  timelines, 218
  track of changes, 220–222

E

EIR. *See* Establishment Inspection Report
Electrocardiogram (ECG), 69
EMA. *See* European Medicines Agency
EPA. *See* US Environmental Protection
    Agency
Establishment Inspection Report (EIR), 41
European Medicines Agency (EMA), 4–5
European Union (EU), 42
Euthanasia, 91

F

Facility inspections, GLP, 40
    See also Good Laboratory Practices (GLPs)
    data audit inspection, 40
    directed inspections, 40
    EIR, 41
    follow-up inspection, 40
    GLP laboratory inspections, 40
    QAU, 40–41
    Warning Letters, 41
FDA. *See* US Food and Drug Administration
Feed consumption, 69, 162
Follow-up inspection, 40

G

GLPs. *See* Good Laboratory Practices
GLP-compliant study
    See also Good Laboratory Practices (GLPs)
    control article, 23
    nonclinical laboratory study, 23
    QAU, 24
    raw data, 24
    sponsor, 23
    study completion date, 24
    Study Director, 24
    study initiation date, 24
    test article, 23
    test system, 23
    testing facility, 24
GLP-compliant study protocol, 149
    body of protocol
        experimental design overview, 154
        objective and purpose, 154
        regulatory compliance, 154
        sponsor, 154
        sponsor representative, 154
        sponsor study number, 153
        study director, 154
        study title, 153
        testing facility, 153
        testing facility study number, 153

compliance with Animal Welfare
    Regulations, 166–167
contaminants, 159
diet, 158–159
experimental design, 159
husbandry, 158–159
number of animals on study, 157
protocol checklist, 171f–172f
protocol finalization
    final nonclinical study protocol, 169
    GLPs, 169–170
records and specimen retention, 166
reviewing draft protocol, 151
    day-by-day study function chart, 152
    experienced laboratories, 151–152
    non-clinical studies, 152
    SOPs, 151
scheduled euthanasia, 164
signatures
    dated signature of study director,
        167
    sponsor approval date, 167
    test facility management, 167–169
statistical methods, 166
study design outline, 160t
study reports, 166
table of contents, 153
test system, 156
    expected age, 157
    expected body weight, 157
    justification, 157
    source, 157
title page, 152
    sponsor, 153
    sponsor study number, 153
    study director, 153
    study title, 152
    testing facility, 153
writing first draft
    ad libitum feeding, 150
    Contracting Scientist, 150
    regulatory, format, and contract
        requirements, 150–151
GMPs. *See* Good Manufacturing Practices
Good Laboratory Practices (GLPs), 1–2, 19,
    103, 207
    auditing checklist, 42
    certifications, 84
    computer operations, 58f
    computer systems, 26
    Contracting Scientist, 21
    CRO and internal company laboratories, 2

Good Laboratory Practices (GLPs) *(Continued)*
  CRO and QA, 20–21
  equipment, 30, 50f–51f
    computer systems, 31
    equipment design, 30
    maintenance and calibration, 31
  EU facility inspections, 42
  facilities, 48f–49f
    animal care facilities, 29–30
    animal supply facilities, 30
    data audit inspection, 40
    directed inspections, 40
    EIR, 41
    floor plan layout for small CRO, 29f
    follow-up inspection, 40
    GLP laboratory inspections, 40
    for handling test and control articles, 30
    inspections, 40
    laboratory operation areas, 30
    QAU, 40–41
    specimen and data storage facilities, 30
    testing facility, 29
    Warning Letters, 41
  facility organization and personnel,
      43f–47f
  good practices or similar terminology, 21
  implementation and requirements, 21
  key personnel and reporting lines, 25f
  laboratory inspections, 40
  nonclinical laboratory study, 36–37
  non-GLP university setting, 2–3
  nonclinical studies
    EPA and EMA, 4–5
    laboratory SOPs, 5–6
    meeting regulatory acceptance, 5
    OECD and QA, 4
    QAU, 5
  OECD, 20
  organization and personnel, 25–26
  protocol, 35–36, 55f–56f
  QAU, 20, 27–28
  QC versus QA, 22
  SOP, 20
  Study Director, 8–9, 27
  study issues, 9
    general, 9–10
    in-life, 13–15
    necropsy, 15–16
    protocol, 10–12
    reporting, 16–17
    test-article, 12–13
  study protocol, 149
  terminology, 24
  test and control articles, 34, 54f, 124
    characterization, 34
    handling, 34–35
    mixtures of articles with carriers, 35
  testing facilities operation, 31, 51f–52f
    animal care, 33–34
    differences from OECD GLPs, 32
    reagents and solutions, 32
    standard operating procedures,
        31–32
  testing facility inspection, 25
  testing facility management, 26–27
  US FDA and OECD GLPs, 22
    electronic records and electronic
        signatures, 23
    RFID, 22
Good Manufacturing Practices (GMPs),
    155–156, 176
*The Guide for the Care and Use of Laboratory
    Animals*, 84–85
  See also Animal welfare
  animal care, 85–86, 89–90
  animal environment, 87–88
  animal facilities, 91
  animal housing, 88–89
  construction, considerations for, 91–92
  general information, 87
  IACUCs, 85
  key functions, 86
  members, 86
  OSHP, 86–87
  performance-based approach, 85
  study design impacting factors, 87
  veterinary care program, 90–91

H

Heating, ventilation, and air conditioning
    (HVAC) system, 92, 90
Histopathology, 32, 165–166
HVAC system. *See* Heating, ventilation, and air
    conditioning (HVAC) system

I

IACUC. *See* Institutional Animal Care and Use
    Committee
ICH. *See* International Conference on
    Harmonisation
ILAR. *See* Institute of Laboratory Animal
    Resources
In-life phase, in GLP animal study, 13–15
In-life study evaluations, 161

See also Good Laboratory Practices (GLPs)
body weight, 162
cageside observations, 162
clinical observations, 162
clinical pathology, 162–163
feed consumption, 162
ophthalmology examinations, 161
physical examinations, 161
unscheduled deaths and moribund animals,
    162
water consumption, 162
Institute of Laboratory Animal Resources
    (ILAR), 84–85
Institutional Animal Care and Use Committee
    (IACUC), 60, 168
key functions, 86
pre-protocol communications, 209
uses for institutional animal care, 85
Institutional Official, 85–86, 92
International Conference on Harmonisation
    (ICH), 60, 121–122

L
Laboratory Animal Technician (LAT), 84
Laboratory Animal Technologist (LATG), 84
Laboratory selection, 103
checklist, 115f–118f
contacting and preliminary screening
    experience, capacity and timing, 106
    meeting with sales representative, 105
    nonclinical laboratories, 104
    protocol, 107
    QAU, 107–108
    reporting, 108
    screening process, 108
    with technical representative, 105–106
    test article and formulations, 106–107
Contracting Scientist, 104
final recommendation, 119f
visiting and auditing new laboratory
    agenda, 111
    animal facilities, 110
    Contracting Scientist's company, 108–109
    critical laboratory functions, 114
    functions and responsibilities, 112
    GLP and animal welfare audits, 110
    information, 109
    outstanding deliverables, 114
    regulatory filings and inspections, 109
    running and monitoring, 110
    SOPs and facility documentation,
        109–110

Study Director, 112–113
working relationship, 111–112
LAT. See Laboratory Animal Technician
LATG. See Laboratory Animal Technologist

M
Master Laboratory Agreement (MLA), 142
Material Transfer Agreement (MTA), 139
Maximum tolerated dose (MTD), 215
Microsoft Excel Spreadsheet, 224–225
Microsoft Word, 220

N
Necropsy, 15–16, 70–71, 190–191
in GLP chronic rodent study, 16
in GLP non-rodent ocular study, 16
No-observed-effect level (NOEL), 215
No-observed-adverse-effect level (NOAEL), 215
No Action Indicated (NAI), 41
Nonclinical study, 1–2, 59
CRO and GLP, 60
necropsies, 70–71
Nonhuman primate diets, 65
Nonhuman primates, 63–64

O
Occupational Safety and Health Program
    (OSHP), 86–87
Ocular tolerance study, 12–13
OECD. See Organisation for Economic
    Cooperation and Development
Office for Laboratory Animal Welfare
    (OLAW), 82–83
Official Action Indicated (OAI), 41
Ophthalmology examinations, 69, 161
Organisation for Economic Cooperation
    and Development (OECD), 4, 20, 60,
    121–122
OSHA. See US Occupational Safety and
    Health Administration
OSHP. See Occupational Safety and Health
    Program

P
Personnel, 26
PHS. See Public Health Service
Plastic cages, 63–64
Postmortem evaluations, 164
    See also GLP-compliant study protocol
    gross necropsy, 164
    histopathology, 165–166
    organ weights, 165, 165t

Pre-protocol communications
  See also GLP-compliant study protocol
  CRO, 209
  customs, 210
  pre-study meeting, 211–212
  protocol development
    draft protocol, 210
    protocol amendments, 211
    sponsor representative, 210
  protocol generation, 209
  study updates, 213
  visiting laboratory
    expected arrival time, 212–213
    procedure performance, 213
    sponsor visiting, 212
Pre-study health assessment, 62–63
Price negotiation, 131
  factors and items, 132
  regulatory compliance and scientific
    conduct, 131–132
Principal investigator, 24–25
Project proposal, 121
  OECD and ICH, 121–122
  price negotiation, 131
    factors and items, 132
    regulatory compliance and scientific
      conduct, 131–132
  study outline template, 132
    animal model, 123, 123f
    basic study design information, 124–125,
      125f
    blank study outline template, 122
    dose groups and timeline of events, 126t
    using non-invasive LifeShirts, 128
    reporting, 135f
    sponsor and laboratory contacts, 122
    starting study and reporting results, 130f
    study title and housing, 134f–135f
    test and control articles, 124, 124f
    variables, 129f
Protocol, 35–36
  amendment, 155, 170
  approval, 150–151
  dosing, 67–68
  GLP and animal welfare, 10
  GLP subchronic large animal study, 11–12
  for nonclinical laboratory study, 36–37
  running nonclinical studies, 11
  specifications, 66–67
  US regulatory submission, 10–11
Protocol title page, 152
  sponsor, 153

  sponsor study number, 153
  study director, 153
  study title, 152
  testing facility, 153
Public Health Service (PHS), 82–83

Q
QA. *See* Quality assurance
QAU. *See* Quality Assurance Unit
QC. *See* Quality control
Quality assurance (QA), 4, 20–22, 215–216
Quality Assurance Unit (QAU), 5, 24, 222
  draft report mail date, 155
  GLP compliance issue, 20
Quality control (QC), 107, 215–216
  versus quality assurance (QA), 22

R
Radio frequency identification (RFID) tags,
  13, 22
Records and reports, 37, 57f
  See also Good Laboratory Practices (GLPs)
  multisite GLP study, 39f
  nonclinical laboratory study results,
    reporting of, 37–38
  OECD multisite studies, 38–39
  records, retention of, 38
  records and data, storage and retrieval of, 38
Report finalization, 237–238
  See also Draft report
  close-out audit, 239–240
  completion, 238
  final report arrives, 238
  regulatory authority, 239
  regulatory requirements, 238–239
RFID tags. *See* Radio frequency identification
  tags

S
Self-monitoring, 92
Signatures
  date of sponsor approval, 167
  dated signature of study director, 167
  test facility management, 167–169
SOPs. *See* Standard Operating Procedures
Sponsor, 23
Sponsor Monitor, 122
Sponsor Representative, 122
Standard Operating Procedures (SOPs), 5–6
  authorization, 31–32
  GLP compliance issue, 20
  protocol amendment, 194

review, 151
stability information, 177–178
Starting study through end of in-life
  See also In-life study evaluations
  acquiring animals, 182
    acclimation period, 183
    animal health, 183
    at laboratory, 182–183
    raising quality animals, 182
  compliance with GLPs, 188
  electronic data capture systems, 189
  first day of dosing, 184–185
    end of day, 186
    laboratory's performance, 185
    right formulations, 185–186
  interim data audit, 190
  necropsy, 190–191
    board certified veterinary pathologist, 191
    experienced technicians, 191
  pre-study meeting, 183–184
    Contracting Scientist participation, 184
    draft protocol, 183–184
    protocol amendment, 184
  protocol amendment, 187–188
  protocol deviations, 186–187
  study updates, 188
  test article and formulations, 181–182
Study communication, 193
  difficult situations, 195
  draft study report, 197
  expectations, 194
  fictional dog study templates, 198–199
    conclusions, 198–199, 202f
    results, 198–199, 201f–202f
    study design, 198–199, 199f–201f
    study timeline, 198–199, 202f
    title slide, 198–199, 199f
  final reports, 197–198
  handling issues, 194
  nonclinical study audit report, 203
  scientist reports contribution, 197
  study contract, 198
  Study Director, 193–194
  study results, 198–199
Study completion date, 24
"Study Contracting and Monitoring,", 1–2
Study design, 59
  See also Animal welfare
  checklist, 73, 74f–79f
  CRO and GLP-compliant study, 60
  general study design issues, 61
    animal housing, 63–64

animal identification, 63
animal procurement and selection, 61–62
assessments, 73
dosing, 67–68
environmental controls, 65–66
in-life evaluations, 68–70
quarantine and pre-study health
    assessment, 62–63
randomization to groups, 66–67
terminal procedures, 70–73
water and feed, 64–65
nonclinical and toxicology studies, 59–60
scientific and practical aspects, 60
Study Director, 8, 24, 193–194
assurance, 27
CROs, 8–9
differences from OECD GLPs, 27
Study initiation date, 24
Study issues, 9
general, 9–10
in-life, 13–15
necropsy, 15–16
protocol, 10–12
reporting, 16–17
test-article, 12–13
Study Monitor, 1–2, 205–206
communication logistics, 207
considerations, 207
  effective working relationship, 207–208
  last-minute changes, 208
  Study Director's role, 208
  study running time, 209
difficulties, 206
expectations setting, 206
expected timelines, 206–207
reporting
  company-specific report template, 215
  documentation, 215–216
  risk assessment benchmarks, 215
  sub-report finalization, 216
Study Director, relationship with, 206
study issues, 213
  study error, 214–215
  test article issues, 213–214
  unexpected or severe toxicity, 214
Study Protocol Design, 149

T

Test and control articles, 155
  See also GLP-compliant study protocol
  administration
    dose levels, 161

Test and control articles *(Continued)*
    duration of dosing, 159
    frequency of dosing, 159
    justification, 161
    procedure, 161
    route of administration, 159
  control article, 155–156
  identity, 155
  reserve sample, 156
Test article, 12, 23, 155
  cellulose derivatives, 12
  considerations
    GLP analytical method, 180
    sampling procedures and schedules, 180
    shipping conditions, 180
    special storage conditions, 179–180
  control article, 156
  disposition, 156
  documentation, 176
  identity, 155
  methylcellulose derivatives, 12
  mixtures, 176–177
    chain-of-custody procedures, 178–179
    CofA specifications, 178
    GLP analytical method, 177
    reserve samples, 179
    stability, 177–178
    variations of concentration and
      uniformity, 177
  properties/characterization, 155–156
  reserve sample, 156
  synthesis and sourcing
    chemical supplier and CofA, 175
    formulations department, 174–175
    sourcing test article, 175
    storage containers, 175
    synthesis procedure, 173–174
  test article/dose formulation, 156
Test Facility Management, 143, 167–169
Test guideline, 154
  See also GLP-compliant study protocol
  alteration of study design, 155
  draft report mail date, 155
  experimental start date, 154
  experimental termination date, 154
  Good Laboratory Practices, 154
  quality assurance, 155
  study initiation date, 154
  submission of study to regulatory authority,
    155
Test system, 23, 156
  expected age, 157

  expected body weight, 157
  justification, 157
  source, 157
Testing facilities operation, 31
  animal care, 33–34
  differences from OECD GLPs, 32
  reagents and solutions, 32
  standard operating procedures, 31–32
Testing facility, 24, 29
  differences from OECD GLPs, 27
  management, 26–27
Toxicology
  GLP, 2
  US FDA, 4–5

U

United States Department of Agriculture
    (USDA), 82
Urinalysis, 70
US Department of Health and Human Services
    (DHHS), 82–83
US Environmental Protection Agency (EPA),
    4–5, 60
US FDA and OECD GLPs, 22–23
  Subpart A, General Provisions, 23–25
    Section 58.3, Definitions, 23–25
    Section 58.3, differences from OECD
      GLPs, 24–25
    Section 58.10, Applicability to Studies
      Performed Under Grants and
      Contracts, 25
    Section 58.15, Inspection of a Testing
      Facility, 25
  Subpart B, Organization and Personnel, 25–28
    Computer Systems, 21 CFR Part 11, 26
    key personnel and reporting lines, 25f
    Section 58.29, Personnel, 26
    Section 58.31, Testing Facility
      Management, 26–27
    Section 58.31, differences from OECD
      GLPs, 27
    Section 58.33, Study Director, 27
    Section 58.33, differences from OECD
      GLPs, 27
    Section 58.35, QAU, 27–28
    Section 58.35, differences from OECD
      GLPs, 28
  Subpart C, Facilities, 29–30
    floor plan layout for small CRO, 29f
    Section 58.41, General, 29
    Section 58.43, Animal Care Facilities,
      29–30

Section 58.45, Animal Supply Facilities, 30
Section 58.47, Facilities for Handling Test and Control Articles, 30
Section 58.49, Laboratory Operation Areas, 30
Section 58.51, Specimen and Data Storage Facilities, 30
Subpart D, Equipment, 30–31
Computer Systems, 21 CFR Part 11, 31
Section 58.61, Equipment Design, 30
Section 58.63, Maintenance and Calibration of Equipment, 31
Subpart E, Testing Facilities Operation, 31–34
Section 58.81, Standard Operating Procedures (SOPs), 31–32
Section 58.81, differences from OECD GLPs, 32
Section 58.83, Reagents and Solutions, 32
Section 58.90, Animal Care, 33–34
Section 58.90, differences from OECD GLPs, 34
Subpart F, Test and Control Articles, 34–35
Section 58.105, Test and Control Article Characterization, 34
Section 58.107, Test and Control Article Handling, 34–35
Section 58.113, Mixtures of Articles with Carriers, 35

Subpart G, Protocol for and Conduct of a Nonclinical Laboratory Study, 35–37
Section 58.120, Protocol, 35–36
Section 58.130, Conduct of a Nonclinical Laboratory Study, 36–37
Subpart J, Records and Reports, 37–39
Section 58.185, Reporting of Nonclinical Laboratory Study Results, 37–38
Section 58.190, Storage and Retrieval of Records and Data, 38
Section 58.195, Retention of Records, 38–39
Section 58.195, OECD multisite studies, 38–39, 39f
US Food and Drug Administration (US FDA), 4, 60, 122, 210
US Occupational Safety and Health Administration (OSHA), 86–87
USDA. See United States Department of Agriculture

V
Ventilation, 88
Veterinary program, 90
Voluntary Action Indicated (VAI), 41

W
Water consumption, 69, 162
Warning Letters, 41
Wire bottom cages, 63–64

Printed and bound by CPI Group (UK) Ltd, Croydon, CR0 4YY

03/10/2024

01040413-0003